THE TERRAFORMING AND COLONIZATION OF MARS

Adding Life to Mars

Charles Joynson

In conservation, the motto should always be 'never say die'.
Gerald Durrell

© 2017 Charles Joynson
All rights reserved.
ISBN: 0995674124
ISBN 13: 9780995674127

CHAPTER 1
INTRODUCTION

When our ancestors first stepped foot on Mars; they began a process which for them was full of danger. Never before had humans attempted to bring a dead planet back to life. Human history is full of stories of people colonising new lands. So the first people to walk out of Africa and the first people into the Americas all took huge risks to reach their destinations. However, Mars was the first place to which people had moved without biology, without air and without water.

This meant that colonisation by small isolated bands of people was impossible and without the support of thousands of scientists and engineers the colonists would not have survived. Therefore the colonisation of Mars was by its complexity an international effort which not even the largest of nations would have managed on their own.

2072
Although we now remember 2072 as the year humans first stepped onto another planet, preparation for the Mars mission had commenced over sixty years before. The challenges which delayed the

launch included the enormous financial cost, the huge risk to life and the vast complexity.

Many of the difficulties stemmed from the need for resupply of the fundamental ingredients needed to sustain life, as before the first Mars landing, people in space had been supplied by scheduled deliveries of such essentials. This meant that water, oxygen, food and medicine could all be replenished at regular intervals. Additionally waste products could be removed before they made the crew sick or damaged the vehicular environment.

But in the case of Mars, frequent resupply missions were not possible. The next ship was expected in twenty years and so everything - absolutely everything - had to be recycled. For example all human waste products had to be reused, with the water recycled for drinking and the solids as a substrate for growing plants and fungi. Anything which could break had to be repairable, and obsolescence was not permissible.

New materials had to be developed which could repair themselves, so that a torn suit could be repaired by allowing the fabric to regrow and a broken camera could have a new part reprinted in a 3D printer and the broken part being liquefied and used to print other objects.

The enormous cost of getting resources to Mars meant that anything which could be sourced on the Red planet would save money, time and potentially lives. The most critical of these resources was water, but many other substances such as clays and metals had to be found so that the colonists could survive.

The choice of settling in a shallow sand filled crater in the Syrtis Major region of Mars meant that the colonists could live in sand tunnels which would protect them from cosmic radiation, solar mass ejections, low temperatures and dust storms. In addition, the ability to extend the tunnel network meant that they could expand the colony without needing additional modules to be supplied from Earth.

There was also a critical formula connecting the number of colonists and their chances of staying alive. Colony success was directly proportional to the crew's gross skillset. The problems they needed to solve broadly fell into three categories; known problems with known solutions; known problems with no known solutions and unknown problems with no known solutions.

This and the financial cost were the key reasons why the mission had to be internationally sponsored, why it needed to involve such a large crew and why it required connection to a wider community of problem solvers.

2073
The early years of the Mars colony were focussed on survival rather than curiosity. Academics and journalists on planet Earth would have liked the colonists to search for life, but as their lives were at risk every second of every day, the location and exploitation of resources were the colony's absolute priorities.

This meant that although sourcing water-ice was top of the colonists' list, comprehensive geological surveys were critical to the location of resources which could increase their survival chances. When a new mineral was discovered, it was analysed and the results were fed to a problem solving super-group on Earth. People who volunteered to be part of the super-group were mostly academics with specialist knowledge of science or engineering. They were sent information the colonists collected about Mars and wrote scientific papers using this data. These papers were forwarded on to the team at Mars control which used them to develop operational plans.

The function of the super-group was split into two parts; the first being forward planning and the second exploitation. Before the colonists had even left Earth the super-group had planned and tested resource utilisation techniques by which critical materials could be collected, processed and utilised. This meant that the crew had enough to keep them busy for the first few months on the planet without the need to go looking for new tasks.

After landing and establishment, the colonists began conducting environmental surveys and the super-group were given information about new minerals and rocks they discovered. For example when the colonists discovered chlorapatite, the super-group were able to advise them how to make bipyridine so that phosphorus could be extracted. As phosphorus is essential to the growth of plants, this was a crucial step in helping the colonists grow their own food.

Each shuttle had its own mass spectrometer and a comprehensive chemistry suite. This meant the user could conduct analyses for the super-group, and follow their instructions to make chemicals to help extract key resources.

Mars' thin atmosphere contained two key gases, nitrogen and carbon dioxide. Both were needed so that the colonists could grow food, and the gases were therefore collected directly from the atmosphere, mixed with oxygen, then warmed and cleaned before being fed into the green boxes which served as underground greenhouses. These were built in the tunnels, and heated and lit by energy from the solar energy collectors on the surface. They used a hydroponic substrate the colonists manufactured from rock granules and water. This was supplemented with mineral nutrient solutions which were circulated through the substrate by pumps.

The atmosphere the crew breathed in the shuttles and tunnels, and the gases in the green boxes were kept separate as plants needed a more tropical and carbon dioxide rich environment than people. This meant that once propagation and planting was finished, the green boxes were sealed and no further human intervention was allowed until harvesting. For repeat cropping plants such as fruit, temporary atmosphere changes were possible during picking, but the green box was then resealed and restored to its previous state as quickly as possible.

Heat emitted from the green boxes was collected and used in a series of fungal hyphae cultures which grew mushrooms on human waste. Any human waste left after the mushrooms had been harvested was used in tunnel construction.

Tunnelling soon became a key element of the operation, and there was great competition between the colonists for ownership of the new chambers. This meant that cements were desperately needed to stabilise, air and heat-seal the tunnel walls. Consequently calcium-rich rocks were greatly sought after, and competition within the super-group to develop dust based glues and cements was intense.

The way the colonists' interaction with the Earth worked was that all communication was split into personal, global community and academic super-group messages. A labelling team created a searchable index of all relevant video, audio and written communication and the super-group academics were then able to follow individual colonists or could search the communication data for key content.

The result of this was that academics could find new knowledge in the communications which no one on Earth had seen before, and could write new scientific papers which were published online and fed back to Mars control. Mars control's operating procedures were also visible to the super-group who could use them as material for further academic papers, thus improving an already good circular process. This feedback advice helped in the creation of strategies which were devised to increase the chance of colony success.

The meeting of tunnel networks in the year 2077 made movement between shuttles far easier and also made the colony feel like a colony rather than just being "three shuttles and some tunnels". The colonists began to believe that they really were colonising a new world and not just fighting for their lives.

After the first five years on the Red planet permanent changes were seen in the colonists' bone mass and muscle density. This meant that return to Earth would have led to serious disability and, even though the colonists took to wearing weighted clothing, any plans for crew swaps were cancelled.

The second Mars mission was launched in 2092 to resupply the colony. The new colonists brought with them renewed enthusiasm, new ideas, new technology and new tools. One new technology which arrived with the resupply was bio-smelting, which used bacteria to leach metals from rocks, avoiding the need for high temperatures and limestone fluxes.

In fact, using living organisms to change one substance into another became a key colony process, with insects and other invertebrates changing inedible plant material into edible protein, algae extracting nitrogen for plant foods, and plants making oxygen for the colonists to breathe.

CHAPTER 2
INTERPLANETARY TRAVEL

In the early years of space travel the time taken to get from Earth to Mars was as much as ten months. The extreme duration of travel at this time exposed crews to solar and cosmic radiation as well as many low gravity health hazards.

The first generation of space travel was powered by chemical rocket propellants which contained either oxygen or a mixture of fuel and an oxidant. These were burnt in a combustion chamber with the high temperature gas jets providing the thrust. These propellants could be solid, liquid or gaseous. Solid propellants were made from granulated mixtures of ammonium nitrate, ammonium perchlorate or potassium nitrate; along with binding agents, plasticisers, stabilizers, and burn rate modifiers. The majority of liquid fuelled rockets used oxygen with kerosene, liquid hydrogen or nitrogen tetroxide. Alternatively some rockets used a single liquid such as hydrogen peroxide, hydrazine or nitrous oxide. Gaseous propellants needed a compressed gas to fuel them. However, the combination of low density gas and the weight of a high pressure vessel made them inefficient and expensive. There were hybrid rockets which had a solid fuel and a liquid or gas oxidiser, making it possible to slow and restart the motor as was possible with liquid fuelled rockets.

Although these chemical propellants enabled humans to leave planet Earth for the first time, they did not, for the most part, allow reusability. As such, a rocket would be designed and built for a single lift following which it was then either burnt up in the atmosphere, exhibited in a museum, lost or scrapped. All of this made getting people and cargo into space expensive, limiting access to state-funded organisations and the biggest of commercial companies.

Electrically powered spacecraft used electrical energy to create thrust. These rockets worked by accelerating matter to high speeds and used much less propellant than chemical rockets. Plasmas were produced by an electric or magnetic field and the plasma was forced out of a nozzle to create thrust. Poor battery performance limited the early versions and these had much less thrust than chemical rockets. This restricted their use to off-planet acceleration and manoeuvring. Later versions of these spacecraft had enhanced fusion power units and used a greater quantity of massive ionised propellants. This increased acceleration and reduced the time taken to make the Mars round trip, but they were still unable to lift loads off planets.

Photopic drives accelerated photons to create thrust. Some used Earth or moon-based lasers, whilst others incorporated a photon generator in the shuttle. However, the low mass of photons made the acceleration insufficient to shorten journey times and they weren't used much.

Magnetic and solar sails were used a few times in early missions to the outer planets. They used the solar wind to push a probe with a sail attached away from the sun. Plasma drives took over from them as they allowed craft to be accelerated around planets to gain additional momentum which was very difficult with sails.

Maglev sky lifts took over the heavy lifting of getting mass into Earth orbit by 2200, with electrically produced plasma being the primary interplanetary drive mechanism until matter modulation took over in 2450.

Matter modulation involves changing one form of matter into another with the intention of forcing as much mass out of a rocket as possible. Normally this meant changing quarks into gravitons or anti-gravitons and electrically accelerating these out in a fast moving stream. This is the primary way we have been getting around the solar system since 2800, at least until we understand more about controlling and bending space.

CHAPTER 3
ATMOSPHERE

2385

To keep surface water on Mars so that the planet could have lakes, rivers and oceans, Mars first needed an atmosphere to prevent water boiling away. This meant that sources of frozen oxygen, nitrogen, carbon dioxide and inert gasses had to be found beyond the gas line which limited where gasses could exist in solid form in the solar system.

The best source for frozen gasses was the Kuiper belt, a long way from either Mars or Earth, necessitating the development of a new generation of autonomous space tugs.

These devices, built in their thousands after 2380, were made up of five components:

1. The engine to take the tug out to the Kuiper belt
2. The control system which located the appropriate frozen gas object and controlled the landing on its shadow side
3. The solar array or fusion power source

4. The laser which turned solids back into gases
5. A funnel which directed gas away from the object into the articulated tube which directed it into space

It was the tube within component in part 5 above which controlled object's trajectory, though of course the type and efficiency of the power source was critical to the amount of gas propelled through the tube. Those tugs which used solar power arrays had one of three types:

1. Folding arrays which folded and unfolded like umbrellas;
2. Unfurling arrays modelled on ferns and other organic structures; and
3. Spray arrays which were painted onto the surface of the gas object.

There were also a number of different approaches to fixing the tug device onto the shadow side of the frozen gas asteroid:

1. Freezing liquid where liquids were used as glues;
2. Hot prongs which involved heated hooks melting their way into the object's surface and then allowing the gas to re-freeze; and
3. Laser drilling which worked in a similar way.

The first gas object to be delivered to Mars was a large chunk of carbon dioxide which was fragmented onto Olympus Mons in 2385. Other mountainous chunks of oxygen and nitrogen followed at regular intervals and were mostly impacted onto the frozen north-pole. This released more carbon dioxide into the atmosphere, thereby warming the planet and the kinetic energy of the impact also had a warming effect.

Thereafter frozen gas objects were brought to Mars on a regular basis so that the atmosphere began to comprise more nitrogen and oxygen than carbon dioxide. 2413 saw the first people breathing outside on Mars without masks or respirators. Water-ice was a regular addition to Mars' atmosphere after 2422 and was mainly impacted into the northern ice cap. The southern pole was avoided so as to prevent flash floods and avalanches. Once the air pressure had reached 17 N/cm^2, most objects were angled to come in tangentially and evaporate in the upper atmosphere. Regular rain storms began to desalinate the soil and wash salt into the northern basin.

To warm the atmosphere, which could not be done by just adding greenhouse gasses, the first of 42 wind towers was put into operation in 2386. These were huge towers built around the equator which resembled lighthouses on Earth but were considerably larger. The base comprised a fusion generator which produced heat and a hollow tower reaching up to the sky above it which served as a cooling system. Air entered the tower 30 meters above the ground through a series of large vents; it was then warmed by internal vanes and was accelerated up toward the top of the tower 300 meters above. Water was sprayed into the tower's interior at various points to add water vapour – another greenhouse gas – to Mars' atmosphere. In this way the average temperature of Mars' equatorial atmosphere was raised from minus 60 degrees Celsius to plus 5.

By 2460 an ozone layer had formed and colonists were able to dispense with hats and creams which had been necessary since 2413. However, the small size of the Red planet and the thickness of the atmosphere meant that erosion became a problem with the solar wind removing unsustainable amounts of the protective layer each year. Therefore, in 2453, a solar shield made up of 9 magnetic field generators was manoeuvred into position to protect Mars from the solar wind.

CHAPTER 4
WATER

There are two kinds of water which occur in frozen form in our solar system. There is the good kind which occurs on planet Earth and which we drink, wash and swim in - and there is the bad kind which is fortunately very rare on Earth. This bad kind or "heavy water" contains deuterium which is hydrogen with a neutron in its nucleus. Normal hydrogen in ordinary water doesn't contain any neutrons.

This excess neutron makes heavy water slightly toxic to animals causing sterility and cell division problems, so heavy water was not the right water for Mars. Heavy water is ancient water which has been bombarded with cosmic rays over very long periods of time, whereas ordinary water was formed more recently.

There was enough water existing on Mars in frozen form before terraforming to fill about one third of the northern basin, but much more water was needed to fill inland basins, lakes, rivers and rain clouds. This early water had flowed on Mars in the distant past because it had been held on the surface by the gasses released by volcanoes. These volcanoes had also heated the atmosphere with the copious amounts of lava they produced. However, the quantity of fossil water was not enough for terraforming.

Therefore, more than twice the volume of all the water that existed on Mars had to be found elsewhere in the solar system and it had to be ordinary water and not the heavy variety. At this time a great deal of survey work had been undertaken in the Asteroid Belt and in the Kuiper Belt. Although ordinary water occurred amongst the asteroids, it was needed in a form which contained as little heavy water as possible and with not too much rock. This meant that the best sources had to be found, prioritised and shipped to Mars.

Centuries later, long after Mars had been terraformed, ways were found of separating the heavy and light waters in asteroids and comets. However, back in 2400 the only answer was to locate objects with as little heavy water as possible.

As regular ice asteroids began arriving on Mars, heavy water sank into the depths of the northern ocean. It was not known at this point whether this would be a problem for the terraforming of Mars or not, but it prompted researchers to investigate if some life forms could succeed in these conditions.

As the northern ice cap was fragmented and evaporated by asteroid impacts; immense landslips and tsunamis ripped chunks out of the southern land mass. The northern ocean went through its birth pains as an immense maelstrom of foaming and roaring water tore around the pole. The increased weight on the northern part of the planet caused earthquakes to further destabilise the continent and even damaged the colony's buildings and tunnels.

Water running off the southern continent added vast quantities of mud and salt to the turbulent ocean meaning life could not prosper until the ocean was cleared.

CHAPTER 5
FOOD

Before Mars was given an atmosphere, the only way to grow food was underground. The danger of cosmic and solar radiation prevented greenhouses being sited on the surface, as did the very low temperatures and the risks that dust storms posed to structures.

Surface greenhouses would have needed to be transparent to visible light, insulating against the low temperatures, and made of materials resistant to radiation. These issues and the hazards associated with growing food on the surface left green boxes as the only option.

Therefore, green-boxes were constructed in the tunnel network and powered by surface solar power and nuclear generators to achieve optimum light, moisture and temperature conditions for food plants. These plants yielded various foods including fruits, leafy vegetables and seeds.

Tube boxes were similarly used to grow various types of algae and seaweeds. This entailed pumping mineral and symbiotic bacteria-enriched waters through a tube network which spiralled around a series of tube lights. The symbiotic bacteria provided the algae with vitamins, and the bacteria benefited from the heat, minerals and water. The algae could be harvested at any stage by introducing filters into the tubes.

The various types of mushrooms grown in the tunnels had the advantage of not needing light. Waste heat from the green-boxes was used to optimise temperatures for the fungi, and human waste products were used as a growth substrate. The only disadvantage was that human waste had the propensity to smell, which meant that the fungal boxes had to be sealed until fungal hyphae had digested most of the waste. After this process had taken place the smell was less intense and mushrooms could be harvested.

A combination of the tube technology and fungal propagation eventually lead to the development of broth foods which became the staple foods for the colonists until an atmosphere on Mars was created.

In the 2220s printed foods became available which were created using molecular level 3D printing machines. Printed food consistency, shape and colour were very flexible and many versions could be created. However, the disadvantage was that all printed foods tasted the same.

The first food plants were grown outside on Mars in 2425. Experience taught the colonists that cultivated forms did not survive well and that it was more efficient to plant wild or semi-wild forms which grew rapidly and were easier to propagate. Successful plants included tomatoes, chillies, grapes, leaf crops, wheats, maizes, peas and beans. Because of the high salinity of the soil, root crops were still too toxic for the colonists to eat, but experiments took place with washed soils in the ground and in tanks.

Once real plant foods became common, printed and broth foods continued to be available and cheap, but real foods tasted far better.

In 2486, chickens, goats, sheep and rabbits were introduced to surface farms. Real rather than synthetic meat was a tantalising luxury previously unknown to the colony. New food plants including coffee, tea, sorghum, oats and sugar cane were cultivated and were recognised as being considerably tastier than printed or broth foods. The first root crops were deemed safe to eat in 2487 as some soils were by then washed clean of salts by years of rainfall.

By 2497 printed foods both on Earth and Mars became better tasting and cheaper than they had been previously. The advantages for Martians were mainly to do with cost. In spite of this, many people continued to prefer real food just because it was real rather than synthetic.

CHAPTER 6
HOME

The definition of Home changed for the colonists as Mars was terraformed from a lifeless desert to the lush environment it is today.

The landing of the first humans in 2072 and the austere nature of the Red planet drove the colonists to live underground to stay alive. Living on the surface at that time would have meant being vulnerable to cosmic radiation, solar flares, dust storms and very low temperatures.

The landing site had been chosen because of its equatorial warmth and the proximity of water-ice and other resources, as well as providing deep stable sand and crater walls which protected the site from Mars' fierce winds. The crater was also large enough for the shuttles to land safely, and the sand was of sufficient density and purity to allow the shuttles to dig themselves down and to begin the tunnelling process.

From the earliest days location beacons were mounted around the colony to help colonists get home when they were lost in Mars' deserts where perspective and distance were illusionary. This meant they could pull up a map showing their location and the direction they should travel to get back to the Home crater.

In 2077 the tunnels met to connect all three shuttles. After this meeting, the tunnel network was gradually expanded outwards so that by 2200 the Home crater was full and no more sand tunnels were possible. In a four places tunnels were hand-drilled through the crater walls. However they were of much harder metamorphic and igneous rocks which meant it took a long time to cut tunnels and great effort was needed to make progress. It would have been easier to blast the rocks using explosives as they would have done on Earth, but on the Red planet this could have caused tunnel collapse and may have also poisoned the air the colony breathed.

By 2093 three properly built surface entry ports had been constructed to allow easier access to the surface. These were made of glass panels and steel girders, and in 2105 the suit ports were moved above ground. This allowed people to experience real Martian daylight in their normal work clothes rather than having to don a space suit and see it through the visor. In 2106 a small café was set up in one access port building within an area enclosed by transparent glass and proved instantly popular.

Although the tunnel network continued to expand outside the Home crater, progress was slow due to the hardness of the local rocks. As Mars was still as susceptible to all the dangers which had forced the colonists underground in the early days, surface living was still very hazardous.

Therefore, in 2300 a start was made on moving the colony to a network of lava tubes just over 100 kilometres from the Home crater. These tubes had the advantage of being considerably larger than the Home crater with space for a city rather than just a colony. Construction took over 110 years to complete but, as building was done in segments with each segment occupying 250 meters of lava tube, families were already able to start living in the "Tube City" as early as the year 2306.

The Tube City was designed like cities on Earth, with offices, homes, research facilities, parks and pod transport networks. The parks included many species of plants; and animals such as small reptiles, birds and mammals were introduced to help people experience nature for the first time.

After 2385 regular supplies of frozen gases were brought from the Kuiper Belt to be impacted into the poles or skimmed into the atmosphere. Hence by 2413 people were able to breathe the air outside on Mars for the first time, unrestricted by the confines of the tubes. In 2425 the first colonists moved out of the colony and began living independently in their own printed mineral homes.

As terraforming proceeded and more frozen gasses were brought to Mars, asteroid impacts triggered earthquakes which caused some Home crater tunnels to collapse, and rainstorms and damp sand produced warping in the surviving tunnel walls. However, and rather fortunately, Tube City endured with only minor damage. Over time what remained of the Home crater tunnels were turned into a museum and Tube City was converted into a hospital and university complex.

By 2450 the colonists were living in a curious mixture of timber and printed stone houses. Many took to gardening and grew fruit, seed bearing trees and shrubs. By 2511 nano-robotic construction had become the main means of creating homes and entailed the planting of a nano-robotic seed which could grow to create a home, an office or a transport module.

Nanos got their resources from metals and other minerals existing naturally in the ground and were able to reproduce and create almost any structure. Once a building was complete they were deactivated so that no further growth interfered with the edifice's function.

Planting a nano seed comprised three stages, programing, planting and feeding. The programming was done by specialist architects, but the planting could be done anywhere and by anyone. This meant that inappropriate planting did take place and disputes arose between neighbours where structures were planted on each other's land. Because of this, land and building planning regulations had had to be written into law, alongside ownership legislation. The last stage of the process – feeding - was only required if there were insufficient supplies of metals or minerals in the seed area. In these cases additional resources had to be added to the ground next to the structure. The nanos then accessed the new resource and finished the building.

CHAPTER 7
HEALTH

The long space flight from Earth meant that many colonists arrived on Mars feeling unwell or weak – or both. The main problems the colonists experienced on the long duration trip were bloating, vision problems, and bone and muscle loss. In addition most experienced some form of jet lag. The best cure for these conditions was to get them working as soon as possible once they had arrived on the Red planet.

Real physical activity seemed to reduce the severity of the ill effects caused by Mars' longer day and low gravity. After the 10 month micro-gravity trip many colonists found Mars' one-third level of gravity a relief and considerably easier to adjust to than the full 1g on Earth. Most new arrivals still experienced difficulties adjusting to the light and dark sequence even if they were working on the surface, as Mars' daylight was weaker and redder than Earth's. This meant the day-night lighting in the tunnels had to work in sync with Mars' days, or sols as they were called, to help people overcome persistent jet lag.

Those who continued to have health problems were encouraged to wear weighted clothes which helped them adjust to the conditions on Mars. Sleep disturbance, depression and bone loss continued to be problems for a large number of colony members and prompted many colonists to take medicines which also helped them adapt to their new surroundings. Task rotation also assisted, so work patterns were cycled so that as far as was possible all the colonists had equal amounts of surface, tunnelling, research and construction work as each other. In the 2550s a symbiont – a microscopic creature which lived in people's blood streams and produced helpful drugs - was developed to keep bone and muscle density high, but some colonists still preferred to use the older remedies.

Another pharmaceutical solution which was brought from Earth on the periodic resupply missions was vitamin D, which the colonists needed to take due to their very limited access to sunlight. All the colonists had been recruited using health as a selection criterion. Those who were selected to join the colony had all been assessed to be in very good health and so very few took any medication besides vitamins and minerals. However, after 2236 when sterilisations were reversed and the first children were born, contraception was added to Mars' pharmacopeia.

Depression and psychological problems were persistent despite all the efforts of the colony's doctors and psychologists to assist those who were unfortunate enough to be affected by mental health issues. The presence of partners and lots of love and kindness were key to recovery; however a few members of the colony who failed to respond to treatment had to be returned to Earth. Amongst the commonly diagnosed causes of the mental health problems was severe separation anxiety due to their isolation from friends and families. This could only be countered by support groups which built kindness and loving networks amongst the colonists.

Another problem was that, although Mars had similar seasonality, day length and planetary tilt, the Martian year was nearly two Earth-years long. This meant that, although the colonists were exposed to a longer summer, the weak sun provided little benefit, and they had to endure the dark red of an eight-month long winter. Light was the best cure for the particularly low mood that nearly all colonists experienced at some point during the winter, necessitating wide spectrum lighting inside the colony with hot spots where people could increase their exposure and raise their moods.

There was some good news for the colony though in that although diabetes and weight control had been anticipated to be problems, the reality was that the control of available foods and the hard physical work most colonists had to do limited weight gain and consequent illness. Until colonists started farming livestock and sugar-rich crops; fats and sugars were not available in shops or restaurants. After farming commenced in the 2480s, some unhealthy products became available, but, as most colonists did hard physical activity every day, obesity and diabetes remained rare.

After 2236 the presence of children on Mars meant that the colonists had to acquire new skills and new people, in order to ensure the health and safety of, and latterly to provide education for, the colonists' offspring. Doctors and nurses had to refresh their paediatric skills and new medical specialists had to be recruited on Earth and transported to Mars. Children born on Mars quickly got used to living underground in airtight chambers. They adapted rapidly to the low gravity and, although there were challenges in finding adequate resources to support the growing needs of the colony, the colony soon became a friendlier and happier place.

Dust continued to be a problem for the colonists and keeping it out of the tunnels and later the Tube City was a major preoccupation. Removing dust was a three-step process:

1. The access ports where colonists re-entered the tunnels had air scrubbers that blew filtered carbon dioxide at people to remove dust;
2. The air lock then removed more dust as they waited for breathable air to be introduced;
3. The suit-ports kept dirt out by unzipping the wearer's suit from the back, allowing the wearer to slip out of the suit through the suit-wall, leaving the suit and all the dust on it still attached to the suit-wall.

However in spite of this process, dust still managed to find its way into the tunnels and had to be regularly removed.

The problem with dust was that, apart from getting into equipment and causing breakdowns, some of it was toxic and led to some members of the colony experiencing allergic reactions. Allergies were always a minor problem and inconvenience to the colonists but, as the years went by they got worse. This was particularly the case in those who were born on Mars, which led to research into the causes being conducted.

Initial attention was focused on dust toxicity, and finding ways to decontaminate environments was given priority. Later, colonist's gut flora came under scrutiny and faecal transplants were performed using microorganisms donated by the healthiest colonists. This helped identify the causes of a few allergies but the majority persisted. Once Mars had an atmosphere and the air became breathable, attention was concentrated on the colonists' immune systems. The lack of exposure to archaea - a primitive kind of bacteria - was found to be the cause. Following the addition of more archaea to the environment, allergies began to vanish.

Although people on Mars lead for the most part long and healthy lives, in the end they all had to die. The first death had occurred in 2121 and, until the air became breathable in 2413, there was no way for bodies to decompose or for them to be cremated. In the early years bodies were placed in shallow graves where they froze and were preserved for later generations to dispose of. In 2174 a city of the dead, modelled on the Petra on Earth was created by carving robots where bodies could be preserved either permanently or for cremation at some future date.

CHAPTER 8

MONEY

Even in the earliest days of the Martian colony, when there had been fewer than 100 colonists, some way of rewarding effort and work was necessary. In those first decades friendship and common enterprise were better at countering depression and separation anxiety than any tangible reward. However, the colonists decided as a group that some freedom of choice incentive for effort and attainment was important when resources were limited. In this way they could avoid over-consumption and share resources fairly. They believed that it only took one person to upset a whole group if greed, gluttony or laziness became issues. This thinking resulted in the introduction of the "Martian dollar" or "M dollar" in 2100, with the supply of money being based on the number of colonists. The overall control of money rested with the colony directorship.

When M dollars were announced there was debate as to whether they should be held only on colony computer systems or if they should have a physical expression in coins and cards. The debate was fairly short-lived as DNA sniffing had become a reliable way of being identified before paying for goods and services on Earth since its introduction in the 2050s, resulting in physical money fast being withdrawn, so there seemed no convincing argument for the colony on Mars to go back to the old ways of doing things.

In its first incarnation the M dollar was classed as a closed currency which meant that no exchange rate or conversion was allowed with Earth currencies. There was also a socialist element, as until colonists had families to support, everyone was paid more or less the same salary. The argument made to support this was that no person could sleep in more than one bed at a time, and as they all had similar workloads, no colonist should eat or drink more than any other. After the first children were born there were additional age dependant amounts paid for each child, which maintained colonist equality. Later when the air outside was breathable some people began living away from the main colony and others started working part time. This changed the equality metric and thereafter salaries depended on the amount and value of work done, much as it did on Earth.

After mining started in earnest on the Red planet, trade necessitated the ending of Mars' closed currency and the introduction of exchange rates with Earth currencies. The closed Martian M Dollar was made fully exchangeable with Earth's currencies in 2445. Most of the mineral and metal mining took place in the Asteroid Belt where microgravity made it cheaper and easier. However, Mars itself had some resources which were not obtainable elsewhere and these could be traded on the markets. Equally, Mars needed Asteroid Belt minerals and metals as well as resources from Earth such as technology, information and knowledge.

After 2410 much of Mars' information need was for the DNA sequences of living and extinct creatures to help make the Red planet habitable. This information, as well as actual living things from Earth, had a cost but, so long as Mars could supply mined resources, the trade was possible. After 2350 when nano mining began and after 2467 when nano building was developed, Mars' help in mineral extraction and structure assembly was much in demand. This meant that the Red planet became a rich planet, able to afford all the new resources it needed.

CHAPTER 9
CHILDREN

The sterilisation of recruits was compulsory for many good reasons in the first 150 years of Martian colonisation. The belief on Earth was that children would interfere with the mission by playing with critical control systems and diverting the attention of the colonists, thereby putting the colony at risk. The authorities on Earth also liked to have complete control over who was working in the Mars project and why they were there. This meant that all voyagers to the Red planet - without exception - were sterilised before take-off from Earth.

For over 100 years the colonists accepted this without complaint, even if the emotional stress was difficult to bear. Life on Mars was difficult and dangerous; it was far too risky for children. The professional attention of the colony was firmly focused on the science of terraforming and so they had no time for distractions like children. Yet without the possibility of children or grandchildren life could - at weak moments - be seen as meaningless. In 2180 voices began to be heard in the colony arguing for reversal of the sterilisations. They said that there was a need for native Martians to take their places in humankind's greatest venture.

Earth's authorities were against this because the provision of human capital was a key way for them to maintain control over Mars and Mars' resources. They believed that, if fewer recruits were sourced from Earth, their control over Mars' natural resources, land and knowledge would be weakened. This was triply true as not only did they supply the people, they also selected and trained them. This meant that they could control the agenda and send pro-Earth colonists to engage in the debate.

This stalemate persisted until 2235 when - without authorisation or official approval - a clinic on Mars began reversing sterilisations for both men and women. The colony directors were ordered by Mars control on Earth to close the clinic and return the doctors to Earth. However, fearing a rebellion, the directors refused to comply. This lead to a breakdown in the relationship between the Martians and the people on Earth which lasted eight months; during which time no ships travelled between the two planets meaning no new recruits, no new supplies and lost trade.

This period of Martian austerity demanded total commitment to recycling and reuse, and required the re-learning of the lessons learnt in the first years of colonisation. Yet, despite its seemingly stronger position, it was Earth which gave way and conceded to Mars' wish to stop colonists from being sterilised prior to their arrival on the Red planet and to permit the reversal of any sterilisations that had not yet been reversed. Earth needed to rebuild its relationship with Mars as the shortage of certain key metals had severely hurt not just critical industries but also many millions of consumers.

The following year the first child was born and Mars needed the 3D printer patterns for nappies, clothing, bedding, toys and many other products. They also needed instructions for making specialist soaps, lotions and foods; all of which had to made locally as soon as possible. As the children grew they also needed teachers to program the teaching computers with Martian skills, and artificial intelligence devices to do the teaching. Learning was primarily achieved interactively in single child booths, meaning that no child was bored or left behind in a class situation.

The education of children on both planets was carried out using small robots called 'Why' machines. These could answer a child's questions all day and all night without tiring. When the child reached the age of five they had an operation to replace the lenses of their eyes with artificial intelligence lenses. These lenses meant the children could interact with a virtual 'Why' machine without needing the real thing.

As the children grew into adolescents the interactive training included learning all the key skills they would need in adult life including maths, physics, chemistry, biology, geology, engineering and history. There was also a strong emphasis on physical skills and sport which involved other young people and made a change from the solitary training they faced the rest of the time.

CHAPTER 10
OCCUPATIONS

The occupations of Martians weren't that different from the types of employment carried out by people on Earth, but there was obviously a lot more emphasis on terraforming and surviving in a harsh environment. So, whilst there were doctors, lawyers and accountants, the majority of colonists were involved in ecology, biology or services to the mining industry.

Mining was one of the most important occupations on Mars, and about a third of the population was involved in this industry in some way. Only a small number were actually mining; the majority were employed in support occupations such as building and servicing mining machines or working on enrichment, in shipping, sales, and many other ancillary trades.

Mining in low gravity was more complex than mining on Earth as any pressure needed to dig or drill a hole had to use three times as much pressure as required on Earth. This meant that mining in its initial form had to involve remote control and autonomous machines heavy enough to cut, drill and move Mars' rocks. This required the machinery to be very large and potentially very dangerous, and of course there were accidents and fatalities.

After 2350 mining machines were replaced by nano robots, which meant that the skills needed for an individual to work in the mining industry changed completely. Nano robots were micro machines which could reproduce themselves, locate resources like key metals and then channel them to a collection point on the surface. The result of the introduction of these machines meant that engineers changed occupation and began researching nano design, development and application. To do this they selected multiple research sites where each nano design could be tested. This became easier as the atmosphere was built and warmed as so that by 2425, when free living was possible, the key mining sites grew to become combined mining and research towns and finally cities.

The enormous amount of research into nanos came to fruition in 2467 when structure nanos were produced for the first time. These could create buildings and components for machines. A single nano could be planted on the resource from which the structure was to be composed, and this would then replicate itself and begin the construction.

Although the first generation of structure nanos were slow, subsequent versions were developed to be faster, meaning that a complete building could be constructed from a single structure nano within two days, and later a complete city block could be achieved in the same period of time.

By 2500 half of Mars' working population was working in nano and associated industries. Nanos were being used for mining and construction across the solar system. This meant that thousands of people were needed to agree the sales of these micro-robots, draw up the associated contracts, make the deliveries and supervise the applications.

The Martians continued to undertake research into the potential of nanos, which had become much feared on Earth due to historical infestations. This meant good ways to deactivate nanos were needed which could quickly turn many thousands of micro robots into frozen, deactivated structures. Other nanos were designed which could change shape when triggered by rain, heat, light and many other stimuli. Nano proofing was also perfected so that nanos would not destroy previously built structures.

Developments in these dynamic nanos led to them being incorporated into the casings of ships and shuttles to protect the crew and cargo against micro-meteorite impacts. The nano layer was engineered to plug any hull breaches so as to prevent the loss of the ship's air.

After the move to Tube City in 2306, the setting up of private businesses was allowed. This meant that entrepreneurs could start businesses which helped the rest of the Martian population. People born on Mars seemed to be better at this and more creative in their business solutions to complex problems than were immigrants. The presence of independent businesses demanded the establishment of support services such as accountants, lawyers and banks.

Food production remained an important part of the Martian occupation matrix. Much of Mars' food continued to be printed using complex 3D printers, but some was still grown in oxygenated liquids. The rest was grown by farmers.

The Terraforming and Colonization of Mars

Many of Mars' academics were racing to research Mars before changed it for ever. As water, oxygen and nitrogen arrived on the planet after the year 2400 there was a great deal of effort to find Martian rather than Earth-derived life. There was also a final effort to understand and map ancient drainage patterns as it was believed that future rivers might follow the same water courses.

CHAPTER 11
LAW

The legal status of each of Mars' colonists was complex because of the way they were recruited and sent to the Red planet. Before they began their epic journeys, each individual was required to sign a document which bound them to demonstrate loyalty and obedience to Mars control on Earth. Their salary was paid into a bank account on Earth and most allowed payments out of this to be made to their families. When Mars created its own currency in 2100, Earth barely noticed as there was no exchange or trade between the two planets.

This situation was maintained until the colony started moving to Tube City in 2306 as, by that time Mars was able to send mineral wealth back to Earth which meant that the Martian colony was able to pay for itself by 2350. Thereafter Mars was a profitable enterprise for Earth which, along with the Asteroid Belt mining enterprises, made many businesses and organisations on Earth a great deal of money.

Then, as private businesses appeared and commerce began in earnest on Mars, a number of influential parties on Earth started to go back on the commitment to pay the colonists for the rest of their lives. When the first colonists had been sent to the Red planet, the countries paying for the venture had not given enough consideration to the costs they would incur over almost 200 years. The result of this was that as soon as the economic situation on Mars became one of profit, the countries which had invested money in the project wanted the businesses on Mars to pay for everything and were reluctant to pay any bills at all.

Tax payers on Earth demanded that Mars should pay for both skills and resources which meant that, if Mars wanted new people they had to recruit and pay for them themselves. This wasn't easy as those Martians being returned to Earth became unwell in Earth's greater gravity. Therefore a dual approach was used which encouraged Martians to have lots of children and sought the assistance of recruitment agencies on Earth to fill key skill gaps.

As Mars became more adept at managing and paying for its own affairs, Martians began to see themselves as being separate from Earth, both economically and legally. On Earth there was talk of charging Mars for 200 years of space travel and colonisation. News of this talk was not well received on Mars and the argument rumbled on until 2440 when newly developed space propulsion engines reduced transport times and costs substantially, making trade much more profitable for both planets, after which it was quietly forgotten.

It was in 2425 that the first colonists moved out of the colony and began living independently. This change from a centralised concentration of people and source of employment made Martians wonder what rights they had under Martian law. A bill of rights guaranteeing the rights of Martians was written and signed by Mars' leaders in 2440. This set out the rights and responsibilities of Martians and was followed in 2445 by an amendment allotting land to those who wanted it. The vast majority of Martians were enthusiastic to receive land, however the process of terraforming made some of these plots untenable as water and life changed Mars' geography forever.

CHAPTER 12

LIFE ON LAND

2410

The primary reason for greening Mars was to make it friendlier to human life. It could have been left in its original state and only changed by the addition of air and water. But, even if the decision had not been taken to add life to Mars, life would have arrived in one form or another in an unplanned manner.

This meant that either the process of adding life was planned or it was unplanned. There would be no pristine, sterile life free Mars as humans carried bacteria, viruses and fungi with them on their rovers, rockets, skins and in their intestines. They might in the future bring other species with them as pets or for experiments. These would get into Mars' environment and could make the planet hostile to humans. Because of this it was considered better to plan the introduction of life rather than to let chance set the agenda.

Following the delivery of water ice and frozen gases to Mars, the atmosphere changed and the planet warmed. This made the possibility of introducing living things to the Martian surface very real. The very first step was to create soil, without which plants couldn't be grown and animals couldn't be fed.

This meant that soil ecosystems needed to be brought from Earth and seeded into Mars' regolith. There were, however, a number of complications which had to be considered before any seeding was done. The extreme diversity and abundance of soil microorganisms meant that the number of bacteria species per gram of Earth soil was commonly in the region of 50,000 and some soils could contain millions of species. Additionally, the numbers of species of protozoa, fungi, nematodes and arthropods were also in the thousands in a single gram in many soils. This meant that sequencing the DNA of all these microorganisms and reprinting them on Mars were beyond human capability in the 2400s, and the only practical approach was to bring soil samples from Earth to be seeded into Mars' regolith.

Frozen soil samples were therefore brought from Earth and propagated in the lab or latterly as the planet warmed, directly into the regolith. For larger more complex species, DNA was re-sequenced from genomic data held on computers and injected into printed germ cells. These were animated by electric stimuli before being incubated in artificial placentas.

No ecosystem on Earth exists in isolation and animals regularly transport nutrients and life from one ecosystem to another. In the first seeding attempts on Mars, life which needed outside support from other ecosystems did not normally survive. For example, lichens from the island of Surtsey which needed seabird droppings in order to grow were not likely to thrive on Mars with no sea and no seabirds.

It was recognised that, while the best seeding approach would be to gradually add complete ecosystems to Mars as temperature and moisture allowed, the first stages needed to include life which could colonise a new environment without inputs from other ecosystems.

This also meant that some parts of an ecosystem had to be seeded onto the Martian surface before others. It therefore followed that soil bacteria and fungi were first to be introduced to the Red Planet and wolves and bears were the last. There was also evidence from previous experience on Earth that seeding was best done in dense groupings, as isolation seemed to make survival less likely.

Another issue which had to be considered before larger plants or animals were introduced was the availability of carbon for life. The main source for plants was carbon dioxide in the atmosphere which meant that the concentrations of this gas had to be constantly monitored, not just to fuel life, but also to ensure it continued to serve its function as a greenhouse gas - keeping Mars warm. Animals and fungi consumed plants so their carbon source was the same, thus insufficient carbon dioxide could delay the terraforming of the planet. When more of this gas was needed it had to be brought to Mars from the Kuiper Belt which took three years to achieve, so it was critical that the timing of the resourcing of the gas was accurately predicted.

There was also an ordering to the ecosystems which were used as sources of life, so in the year 2410 the surface of Mars was still as cold as Antarctica, which meant that whilst life from Earth's southern continent might survive, life from warmer places would not.

This also meant that as Mars warmed, life from warmer Earth ecosystems could be seeded. Hence when Mars' average equatorial temperature rose to five degrees Celsius, previously added Antarctic life began to move away from Mars' equator to colder places like Mars' South Pole and the slopes of Olympus Mons.

In selecting the places on Earth from which life could be sourced, three things needed to be considered. The first was the rate at which Mars' temperature and water availability would rise. The second was a series of Earth ecosystem types called Koppen climate classifications, eleven of which would eventually apply to Mars. And the third was the need for a volcanic terrain within each Koppen region as a source of the critical microorganisms which could create living soil out of Mars' inorganic volcanic regolith.

The introduction of multiple life forms from a niche within an ecosystem was believed to be more likely to lead to species survival. The best reason for introducing lots of different types of life was there was still great ignorance about what would survive and why.

The reason for the presence of a volcano or volcanic terrain in each source zone was that, although Mars' surface had been weathered by dust storms for billions of years, it had not been subject to water weathering since its very earliest days. This meant that the surface had more chemical similarities with fresh volcanic terrain than with a billion year old landscape on Earth.

Each Koppen zone was used as a source for life which could be added to Mars. Each ecosystem type was split into succession stages based on temperature, so the soil microorganisms at the top of the volcano were classified as stage 1, lichens and mosses growing further down were stage 2, small plants growing above the tree line were graded stage 3 and so on, down to the forest in the valleys below.

The seeding was done along a line which extended 200 kilometres both east and west of the colony's Home crater and appropriate seeding points were chosen based on local conditions such as moisture, temperature, mineralogy and regolith structure.

This line, called the seed line, allowed the spread of life to be monitored both to the north and south of the line. The seed line also had the added advantage of avoiding hyper-toxic patches which would have killed anything planted on them. Normally, seeding from different Earth sources was kept widely separated so, when two similar ecosystems came from different continents, one would be seeded to the east and the other to the west of the colony. This allowed success to be closely monitored without interaction taking place between similar species from different continents.

The reason for all the monitoring was not just to see what worked well on Mars and why, but additionally to learn the best approaches to take when one day humans began seeding more planets. It was also believed that new sorts of ecosystems could be created on Mars which included life from a number of different Earth sources. However, seeding from individual sources allowed the researchers to understand and monitor ecosystem loyalty and cross fertilisation.

12.1 Antarctica
The very first Koppen source for seeding onto Mars' seed line was from Antarctica, which with an average equatorial temperature of -15 degrees Celsius, was as cold as Earth could get. Furthermore Antarctic ecosystems normally contained very little available water,

making them ideal for Mars in 2410. The life extracted from Antarctica was sourced from the central and peninsula regions and was transported to Mars in frozen form.

All the Antarctic life was seeded into a one hundred square meter area situated one kilometre west of the colony, and included algae and bacteria which could take nitrogen out of the atmosphere and add it to the soil. Most of the life from the Antarctic Peninsula was seeded in the shelter of a low bluff which protected it from the wind; life from the centre of the Antarctic continent was seeded into the flat unprotected part of the seeding patch.

The introduction of peninsula life consisted of six stages with each stage being determined based on temperature and moisture requirements:

1. Volcanic microorganisms
2. Soil algae, fungi and bacteria
3. Lichens which lived inside rocks
4. Lichens, mosses and liverworts which lived on the surface of rocks
5. Invertebrates which survived by scavenging the dead cells of the previous stages
6. Grasses and shrubs

With the extremely cold conditions, gene-editing was undertaken to improve the ability of early species to survive conditions which were unfriendly to most forms of life. This included improving the ability of microorganisms and lichens to extract nitrogen from the atmosphere. The genetic editing was done on Mars, but the DNA editing instructions were sent from Earth. Of all the lichens which were seeded in the Antarctic phase - three component lichens

- which included cyanobacteria, algae and fungi - were favoured as they were more efficient at fixing nitrogen and were better at surviving the hostile conditions.

Microorganisms were cultured and buried about three centimetres deep in the regolith of the seeding area. Lichens were cultured in warm, damp and well-lit green-boxes which allowed the small lichens samples to grow rapidly so they could be cut into strips for the next phase. Stone slabs about ten centimetres across were then brought into the tunnel labs, placed in green-boxes and a small piece of lichen placed on each. Once the lichen had anchored onto the slab in the warm and wet conditions, temperature and moisture were gradually reduced to match the conditions in the proposed seeding area. The stones with their attached lichens were then placed a metre apart in the seeding area and the growth rates monitored.

Bringing live samples from Earth was always problematic as, whether frozen or dried, not all survived. Plants tended to have the best rate of surviving the journey to Mars as seeds and spores were better able to cope with cold dry periods than live samples. Plant seeds were grown in the green-boxes, in high carbon dioxide concentrated atmospheres and with more blue light than normal, until they could be cultured using micro-propagation tissue culture techniques to create many plants from a single seedling. The cultured seedlings were then acclimatized for three months in the green-boxes and mycorrhizal fungi added to improve their nutrient extraction abilities before being planted in the seeding areas.

The introduction of grasses and shrubs from the peninsula, as well as the introduction of such life in the early stages in other zones, was encouraged by the addition of artificial nutrients to the developing soil. These nutrients contained nitrogen, phosphorus, potassium, magnesium, iron and manganese which speeded up the growth of the seeded ecosystems. Central life, which included life from other central Antarctic regions, was represented by just stages 1, 2 and 3.

The stage 1 microscopic life included soil bacteria, viruses and algae from the colder parts of an active volcano called Mount Erebus. This was important as introducing volcanic soil organisms from Earth was the best way to begin the chemical weathering of Martian regolith and the turning of rock into soil.

One addition which took place as a part of the stage 1 seeding was the addition of bacteria which could extract calcium from rocks. This was critical to concentrating calcium carbonate, without which life would not have prospered on Mars.

12.2 High Arctic Tundra
By 2413 there was some leakage of species from the Antarctic plot into the surrounding area. The planet's equatorial temperature had risen to an average of approximately -5 degrees Celsius, the air pressure was now similar to Earth's and there was more rain and therefore more moisture in the soil.

This meant that the first stage of tundra seeding was initiated on the seed line centred on a point one kilometre east of the colony, with each of five one hundred metre square plots representing a tundra source. Hence there was a plot for each of the Arctic regions; Ellesmere Island, northern Greenland, Svalbard, Franz Josef Land and Severnaya Zemlya. As they were all similar, it was not thought to be necessary to separate them by large distances, but stones and rocks were used to delineate the patches and monitor any escapees.

The Terraforming and Colonization of Mars

Like the Antarctic seeding, the tundra seeding was done in stages. However recently created volcanic soils were absent from the five source zones, which meant that the first stage in the normal seeding sequence was absent. There were nevertheless the Ellesmere Island Volcanics, which at 90 million years old were much older than the seeding team would have liked. But, in spite of their age, they were used as they contained enough microorganisms to be of use in terraforming Martian regolith.

The timing of each successive stage was determined by the growth of the previous, so each stage was dependant on the last for its own survival. The first seeding took place in 2413 and the last stage in 2570. The stages were summarised as:

1. Volcanic microorganisms
2. Soil microorganisms and invertebrates
3. Lichens and mosses
4. Prostrate willows, dwarf birch, grasses, poppies, rushes and saxifrages
5. Alders, blueberries, cranberries, crowberries and heathers
6. Insects and spiders
7. Lemmings and shrews
8. Finches, buntings, and sparrows
9. Ptarmigan, arctic hares, caribou and musk oxen
10. Arctic fox and snowy owl

Decomposition fungi and slime moulds were added during the stage 6 introductions to help to digest dead wood and add carbon to the developing soil. These were supplemented in all the subsequent seeding phases.

The seeds for Mars' first trees came from the Botanic University of Kew in England which held seeds for millions of plant species and could guarantee sterility and the absence of pathogens. They were also able to supply seeds which were genetically diverse enough to be able to withstand high salt and heavy metal concentrations. This meant that if the seeding zone had patches of toxic regolith, the diversity of the seeds was variable enough to mean that some of the seeds would survive and grow into healthy plants.

The last two stages were only attempted once tundra life had spread into an area fifty times their seeding size. This meant that there was sufficient grazing for small herds of larger herbivores, and the stage 10 carnivores were introduced five years after stage 9 to reduce the competition for grazing between stages 7 and 9.

As temperatures rose over the centuries, tundra life migrated away from the equator toward colder latitudes and the slopes of Mars' extinct volcanoes. As this happened, seeding had to follow it, so the seeding of stages 7 to 10 were done well away from the seed line and in a number of different localities.

While it might have been possible to bring lemmings from Earth in a live state, transporting larger animals would have been extremely difficult. The breeding of advanced animals was done using a combination of DNA re-sequencing, stem cell printing and robotic mothers. The robotic mothers were advanced enough to be able to feed the young and to teach them the basics of survival.

12.3 Low Arctic Taiga
By 2430 Mars' equatorial average temperature had risen to zero degrees Celsius. This meant that the next seeding campaign could be initiated on the seed line centred on a point five kilometres west of the colony.

The Terraforming and Colonization of Mars

The two source zones, Russia and North America were similar enough in their diversity to be separated by just one kilometre of unseeded regolith. The stages in the 2430 seeding campaign were:

1. Volcanic microorganisms
2. Soil microorganisms and invertebrates
3. Lichens and mosses
4. Spruce, larch, pines, birch, cottonwoods, alders, willow and poplars
5. Insects and spiders
6. Beavers, squirrels, porcupine and voles
7. Microorganisms and invertebrates in rivers and lakes
8. Fish in rivers and lakes
9. Snakes, salamanders and frogs,
10. Moose, caribou, elk, deer and bison.
11. Gyrfalcons, ravens, wolverines, ermines, martins, weasels, otters, mink, badgers, eagles and buzzards

The first stage volcanic soil microorganisms included bacteria, fungi and viruses from volcanoes in the Kamchatka peninsula and from the southern Alaskan volcanic region.

The stage 1 seeding was done in 2430 and stage 11, the final stage, was undertaken in 2676. The 246 year duration of seeding was in part due to stages 6, 7 and 8 needing the presence of lakes and rivers - which were both common by 2650 due to regular rainfall, but this requirement meant that stages 9, 10 and 11 could not commence until this time.

Due to Mars' low gravity, stage 4 trees grew quicker and taller than expected. Researchers had suggested this was likely to be the case over two centuries before, and this trend was mirrored in other seeding zones.

The addition of large predators such as wolves, snow leopards and Siberian tigers was hotly debated at the time due to the large and dangerous nature of the animals under consideration. A decision was therefore taken to delay their introduction until prey species were present in sufficient numbers to warrant predator introduction.

In the rearing of almost all large mammals and birds, learning to be afraid of humans was part of the training given to them by their robotic mothers. This helped to keep them away from human colonies and reduced the number of negative animal-human interactions to an absolute minimum.

12.4 Mount Fuji

The next ecosystem to be seeded onto the seed line was from Japan and was added five kilometres east of the colony. The seeding was started in 2440 and finished in 2654, and was the first to include dense deciduous forests which managed to survive despite the low Martian temperatures at the time.

It was different enough from both the preceding Taiga and from the forests which were seeded after 2600 for a separation between the previously seeded areas and this Japanese ecosystem to be important. This meant that the researchers could watch both the expansion of the Fuji ecosystem and the cross-fertilisation between it and other forest biomes.

The 12 stages of this seeding campaign comprised:

1. Volcanic microorganisms
2. Soil microorganisms and invertebrates
3. Lichens, mosses and ferns
4. Small plants and heaths

5. Birch and larch
6. Firs, oaks and hemlocks
7. Beech, maples, cypresses, pines and bamboos
8. Insects and spiders
9. Pigeons, thrushes, warblers, ptarmigan and nutcrackers
10. Squirrels, rabbits, macaques, bats, mountain goats, wild pigs and deer
11. Owls, eagles and foxes
12. Bears

In the Mount Fuji region bats commonly lived in lava tubes. However on Mars' seed line there were no lava tubes, so slabs of rock were moved to create artificial caves which the bats found equally hospitable.

The forest floors at the foot of Mount Fuji were composed of compacted volcanic rock, demonstrating that thick vegetation could grow even over the densest regolith. Both Japan and Mars experienced the same dense growth of trees despite prior research which had suggested that the Mars regolith shouldn't have been able to support such growth. It was later recognised that the cause of this density and the ability of trees to grow in such profusion on such difficult soil was due to the ability of the soil microorganisms to extract nutrients and critical elements from both rock and the atmosphere.

The addition of Japanese Asian black bears to the ecosystem in 2654 was the first time a large predator had been added to live amongst Mars' growing vegetation. The introduction of these bears was beneficial as they were mostly herbivorous and ate leaves and nuts from a variety of different plants and were good at excavating for plant roots, invertebrates and digging winter hibernation dens- which helped to further turn regolith into soil.

12.5 Surtsey / Tierra del Fuego

Because of the lack of volcanic terrain in the Tierra del Fuego islands, organisms from the island of Surtsey in the northern hemisphere were used in 2450 to prepare the soil and start the succession.

Both territories have an average temperature of about five degrees Celsius which is lowered by wind and rain.

Seeding was done sixteen kilometres west of the colony and this distance was thought to be sufficient to keep it away from other biomes. As was standard procedure, the seeding campaign utilising sources from the islands of Surtsey and Tierra del Fuego was split into stages:

Surtsey

1. Soil microorganisms and invertebrates
2. Lichens and mosses
3. Small plants

Tierra del Fuego

4. Soil microorganisms and invertebrates
5. Grasses
6. Temperate Nothofagus forests with some conifers and ferns
7. Insects
8. Microorganisms and invertebrates in rivers and lakes
9. Fish in rivers and lakes
10. Geese, ducks, herons and parakeets.
11. Lamas and guanaco
12. Pumas and eagles

One issue which needed discussing with reference to each source zone was what to do about invasive species. In Tierra del Fuego, beavers had been introduced in the early twentieth century and since then had damaged or eliminated many native species. Concerns were raised that, if beavers were introduced to the Tierra del Fuego seeding zone on Mars, they may well do the same to other species which had already been introduced on Mars. Conversely, the alternative view point put forward was that beavers were a highly successful species and should be introduced to help terraform the regolith.

Some clarity was introduced to this dilemma by the statement that only native species should be added to each ecosystem. In this way the seeding expert biologists could monitor the success of the whole ecosystem rather than just looking at individual species within it.

In time the beavers did spread out from their place of introduction in other ecosystems, but, although they did spread out from this point, there were many parts of Mars which they failed to colonise. This may have been due to soil chemistry, temperature, or the local ecosystem. The biologists suggested that some trees had tastier bark and leaves than others.

12.6 Mount St Helens

It took another fifty years for Mars' average equatorial temperature to rise to ten degrees Celsius. At this point the ecosystem from Mount St Helens was sampled and seeding began sixteen kilometres east of the colony in 2460. Even though the temperature was warmer than it was at the time of the initiation of seeding from other zones, species establishment took a long time and only the close grouping of seeded plants speeded this up.

The stages of seeding of life sourced from Mount St Helens were:

1. Volcanic microorganisms
2. Soil microorganisms and invertebrates
3. Lichens and mosses
4. Grasses, lupins and other small shrubs
5. Insects and spiders
6. Willows, hemlock and alders
7. Firs and pines
8. Lake and stream microorganisms and invertebrates
9. Lake and stream fish
10. Chipmunks, gophers, ground squirrels, mice, voles, bats and shrews
11. Sparrows, warblers, flycatchers and other birds
12. Frogs, toads, salamanders, and newts
13. Beavers
14. Moose, caribou, elk, deer and bison
15. Weasels, eagles and owls

Some of the Lupins in stage 4 were good at adding nitrogen to the soil, which was important for the establishment of grasses and subsequent stage plants. The positioning of early stage seeded life was important as having been sown in sun-lit cracks and gullies, the seeds were protected from the wind and this encouraged growth, presumably due to the higher temperatures to which they were consequently exposed.

12.7 Eastern United States Temperate Broadleaved Forests
By 2470 temperatures at Mars' equator were warm enough to allow the planting of broadleaved tree species. The average temperature had risen to the extent that people could swim in the lakes in the summer and could also comfortably hike - even in the winter.

Therefore, six new seeding zones were chosen with the first being seeded with an ecosystem from the Eastern United States.

The seeding of these Temperate Broadleaved Forests was not confined to the seed line but, with better transport options then available, the seeds were scattered around Mars' equator. All six zones were seeded as much as was possible at the same time and in parallel with each other.

1. Volcanic microorganisms
2. Soil microorganisms and invertebrates
3. Lichens and mosses
4. Wildflowers, herbaceous and low growing woody plants
5. Oaks, beech, maple, birch, chestnut, ash, aspen, basswood, cherry, elm, hemlock and walnut
6. Wild grape, poison ivy, and Virginia creeper
7. Insects including butterflies, moths and beetles
8. Mice, voles, squirrels, shrews and moles, bats
9. Lake and stream microorganisms and invertebrates
10. Lake and stream fish
11. Jays, robins, nuthatches, woodpeckers, hummingbirds and wild turkey
12. Martens, armadillos, opossums, beavers, weasels, skunks, foxes and raccoons
13. Frogs, toads, salamanders, newts and snakes
14. Deer
15. Hawks and eagles
16. black bear, bobcats, wolves, mountain lions, and coyotes

The Chestnut Blight, Dutch Elm and Ash Die Back and other tree diseases were screened out by sterilising the seeds before their transportation to Mars. This avoided the catastrophic loss of trees which had happened around the world in previous centuries.

12.8 European Temperate Broadleaved Forests

Europe's forest succession was similar in many respects to that found in the United States. However although the seeding stages were similar, the species were different. The volcanic primary succession species to colonise bare regolith came mainly from New Zealand and from China - but some samples were also brought from volcanic landscapes in France's Central Massif. These species were older than the seeding teams would have liked but did provide useful additions to the list of microorganisms.

1. Volcanic microorganisms
2. Soil microorganisms and invertebrates
3. Lichens and mosses
4. Grasses, herbaceous and flowering plants
5. Insects
6. Hazel, Birch and other colonising trees
7. Oak, maple, beech, elm, ash and willow
8. Lake and stream microorganisms and invertebrates
9. Lake and stream fish
10. Squirrels, mice, voles, bats and shrews
11. Sparrows, woodpeckers and other birds
12. Frogs, toads, salamanders, and newts
13. Beavers and wild pigs
14. Moose, deer, mountain goats and European bison
15. Weasels, otters, martins, eagles, ravens and owls
16. Wolves, bears and lynx

Small numbers of large predators had been introduced as parts of a number of ecosystem introductions, and their small populations were considered to be too small to control herbivores by most biologists. Yet the delay allowed prey species to breed up to sufficient density to keep the predators healthy.

12.9 New Zealand Temperate Broadleaved Forests

New Zealand's primary succession species came from its Volcanic Plateau on the north island, much of it from Mount Tarawera.

Some of the first plants introduced as part of the New Zealand succession were New Zealand brooms which were efficient at adding atmospheric nitrogen to the soil by growing root nodules which contained nitrogen-fixing bacteria.

1. Volcanic microorganisms
2. Soil microorganisms and invertebrates
3. Lichens and mosses
4. Herbs, brooms, gorse, tree lupins, bracken and tree ferns and grasses
5. Insects
6. Pōhutukawa, kānuka and mānuka trees
7. Beech, kauri, kāmahi, podocarp, tōtara, miro and tawa trees
8. Birds, such as saddlebacks, tuis, fantails and bellbirds.
9. Skinks, geckos, tuatara, frogs and bats
10. Owls and moas.
11. Haast's eagle

The big problem with species of plants and animals from New Zealand was that many of the natives were extinct and many of the available species like wombats were exotic introductions. This meant that decision had to be made concerning both the resuscitation of extinct species and the exclusion of most exotics, and once decisions were made they had to be put into practice.

The process of recreating a life form's DNA was a highly complex process as, for the most part, any existing DNA was highly fragmented. The result of this was that the process of recreating a huge jigsaw of chromosomal and mitochondrial DNA was highly complex and, without the assistance of artificially intelligent sequencing machines, a life form's complete DNA might have taken hundreds of years to piece together.

There was also the problem of where to get an extinct animal's gut and skin flora from. Principally this meant finding similar species and sampling the microorganisms in their guts and on their skins. Sometimes this worked, at other times it didn't. There were three sources from which gut and skin biomes could be sampled:

1. Related creatures living in similar environments
2. Related creatures living in different environments
3. Unrelated creatures living in similar environments

Sometimes all three sourcing techniques were used at once.

Another debate arose as to whether to introduce gorse, a European invasive plant – it was decided to include this in the Mars seeding as it was good at colonising bare landscapes and at sheltering tree seedlings.

A further debate centred on the extinct Haast's eagle, which had been the top predator until the Māori arrived. Four de-extinct species of Moas were added toward the end of the seeding process and the Haast's eagle was introduced once the Moas were breeding successfully.

12.10 Tasmanian Temperate Broadleaved Forests
Tasmanian cool temperate rainforest from the Savage River National Park was the main sampled ecosystem used when introducing life from this area to Mars. Microorganisms for the

Tasmanian seeding were sourced from the parts of the Savage River Park underlain by basalt.

Venomous snakes such as the Tasmanian tiger snake, and poisonous spiders were not introduced, mainly due to the colonist's objections. It was felt that, when ecosystems started to merge, there would be enough predators to control Tasmanian animal populations without the need to introduce others that would put the colonists' own lives at risk.

Other exclusions from the Tasmanian seeding campaign included a number of exotic plants such as blackberries, and animals such as dingoes - which could have spread across other Martian ecosystems.

1. Volcanic microorganisms
2. Soil microorganisms and invertebrates
3. Mosses, liverworts, lichen and fungi.
4. Wet scrub and buttongrass
5. Insects and spiders
6. Small broadleaved trees, vines, ferns and shrubs
7. Myrtle-beech, eucalyptus and pencil pine trees
8. River microorganisms and invertebrates
9. River crustaceans and fish
10. Frogs and reptiles
11. Bats and marsupial mice
12. Birds including parrots, emus, pheasants, swifts, ducks and geese
13. Wombats, pademelons, echidnas, quolls and platypus
14. Thylacines, Tasmanian devils, eagles and goshawks

De-extinct marsupial wolves or thylacines were introduced as the only Australian apex predator. However, when other predators appeared in the Australian seeding zone, many of the thylacines moved to other zones as well.

The last factor which was identified as an issue was that a key management tool in the Tasmanian zone was fire. This meant that a burning cycle had to be established, without which tree seeds would not germinate and the biome would not expand.

12.11 Chinese Temperate Broadleaved Forests
The Eastern Himalayan Broadleaf & Conifer Forests were used as the source for most of the Chinese species to be seeded. The only exceptions were those species that formed the first three stages of the Chinese seeding campaign which came from the Kunlun Volcanic Group at the far western end of the Himalayan chain. The seeding campaign for this area comprised:

1. Volcanic microorganisms
2. Soil microorganisms and invertebrates
3. Lichens and mosses
4. Herbs, bamboo and grasses
5. Insects and spiders
6. Trees including oaks, rhododendrons, maples, magnolias and acers
7. Birds such as flycatchers, pheasants, hornbills, herons, quail, thrushes and leaf-warblers
8. Langurs, red pandas, flying squirrels, muntjac, macaques and bats
9. Musk deer, Himalayan tahr, Bactrian camels and the takin goat-antelope
10. Civets, martens, owls and eagles
11. Clouded leopards, Bengal tigers and bears

Although not a part of this specific ecosystem, giant pandas were introduced in stage 11 simply because people on Mars wanted to see them. Their introduction was made in spite of the doubts of Mars' biologists as to whether they would survive because of their reliance on a single food plant.

The large stage 11 predators were not introduced until the 2850s, which gave prey species in this and other eco-zones sufficient time to reach optimum populations. Snow leopards and Siberian tigers were introduced into other ecoregions at the same time to control the expanding deer populations.

Tiger sized predators which were introduced as part of the Temperate Broadleaved Forest introductions could have been seeded as part of the Mount St Helens or even the Low Arctic Taiga ecosystems. However their introduction was delayed to ensure there was a sufficiently large prey population to ensure their survival. Once it was recognised that there was plenty of food for all these species and little risk that they would be a problem for Mars' human population, their introduction began. The last phase of their introduction is happening now and, although the planetary large predator population is low, it is likely to rise as populations grow and ecosystems continue to merge.

12.12 Land Life Summary

In retrospect, the seeding sites on the seed line should have been more widely separated, but 2410 the colonists had no easy way to travel long distances away from the colony. One factor which encouraged the growth and spread of seeded life was the diversity of seeding. This meant that collecting seeds from other areas close to the collection zone was a very positive step in encouraging dispersal. So, in the case of Mount St Helens, life was also collected from other volcanic environments within one hundred kilometers of the central peak.

Early establishment of stages in any zone was very slow as pollinators and seed dispersers were absent. Step changes in zone expansion happened when these cornerstone species were introduced. Thus adding pollinators such as bumblebees allowed the colonized zone to expand four times its original size, and the addition of seed dispersers like birds, squirrels and herbivores helped the acceleration and continued growth of the zone.

The succession specialists who planned the sourcing and seeding frequently broke their own rules and went looking for species from outside the source zone. This was due to the importance of nitrogen and other elements in the soil in allowing the growth of plants, and meant that species such as myrica faya from Hawaii and gorse from Europe were added to the temperate broadleaved succession to increase the rate at which nitrogen was added to the soil and thereby the time it took for a complete forest to appear.

This was also the reason for the compilation of the 'Nitrogen Fixers List' which included species which could take nitrogen out of the air and add it to the soil. The list was principally made up of bacteria, but also included plants and insects which had a close relationship with these microorganisms and therefore made the process more efficient.

Ecosystem mixing started almost as soon as the first species were added to the High Arctic Tundra in 2413. Now very little remains of the original seeded ecosystems, as most species have moved to new locations which were more accommodating to their manner of life. The result of this ecosystem mixing meant that the tundra ecosystem split into a number of zones with one part moving to be close to the southern ice cap, another part to fringe the northern ocean, and the rest onto the Tharsis Mountains. In the mountains they grew high up in the dry regions, above the rain forests where air was forced up by local weather patterns.

These movements were for the most part gradual and happened as the climate and local conditions changed. In the case of plants, this may have been because their seeds were moved by other agents such as wind, birds or herbivores, but may have also been because the seeds which germinated and prospered created transition corridors based on local availability of nutrients, moisture and temperature.

The movement of animals was done primarily in a search for food, so good grazing attracted deer, and lots of deer attracted wolves. Any thoughts about preventing the mixing or retrospective un-mixing of ecosystems did not last long. The movement of large predators like tigers and leopards was monitored using implants and satellite location. Fortunately though their "fear of humans training" and their preference for cold environments were effective at keeping them away from human habitation and minimised conflicts.

There has, however, been an attrition of species, as local extinctions modified the seeded biomes, resulting in about a seventy percent of all life seeded onto Mars failing to survive. This survival ratio is greater than the biologists expected, but it had been very difficult to forecast.

One constant problem which took centuries to change was the dust storms which had repeatedly buried the growing plants.

Sometimes the plants were found to be still alive a year later when wind and rain exposed them again, but growth through dust was slow.

The use of robotic mothers to teach the first generation of most advanced animals important life skills was a major part of the seeding initiative for hundreds of years. This was critical as, without this training, the young would not have learnt key nurturing, feeding, hunting, cleaning and cooperating skills. Because of this the autonomous mothers had to be continually updated with the lessons learned through the experience of naturalists who had studied the species for hundreds of years on Earth.

Learning to recognise predators was another form of training which was used to teach first generation young animals what to fear and what to run from. In general they learnt from the behaviour of their robotic mothers as to how they should behave when they recognised a predator. The use of robotic predators – which did the young no harm - had to be repeated until the young animals were able to recognise and flee from them, after which real predators could be introduced.

In some cases, complete robotic ecosystems had to be built to ensure the survival of some species such as Japanese and Chinese monkeys. In this way they had better chances of survival in the complex and evolving Martian ecosystem. Many other forest ecosystems were considered for introduction, but the majority were too warm for Mars and their species would not have survived Mars' cold temperatures.

There was a great deal of need at this time for access to DNA, seed and frozen tissue banks, which were used to recreate long extinct forms of life to help with the terraforming of the Red planet.

However, many tissue resources were in poor condition and needed work to be put into a form that could be used. Where tissues stored on Earth needed to be re-sequenced, there were enormous difficulties with the fragmented nature of the data collected. This applied principally to tissues stored in formaldehyde and in dry states. This meant that the fragments needed to be put back together, a job which would have taken humans decades. For that reason artificial intelligence was used to reassemble the codes, a task which the computers found easy and were able to do quickly and at low cost.

The introduction of mammoths, re-engineered from frozen DNA found in Russia over the previous half millennium, meant that grasslands could exist on Mars. Mastodons had been the original leaf eaters, but the absence of mastodon genetic code meant that modifications had to be made to Mammoth DNA. Both the DNA of their intestinal tracts and of their microbiota had to be changed to allow them to eat tree leaves as well as grass, and this caused mammoths to evolve into two species; grass-eaters and leaf-eaters. Without mammoths pushing over trees to get at the foliage, forests would have taken over most of the planet.

Another de-extinction ecosystem addition which happened at the same time as the mammoths re-creation was that of woolly rhinos which were restored from the DNA in frozen specimens found in Siberian mines. They had been grass and sedge eaters, so their intestinal microorganisms had to be sourced to allow them to re-fill this niche on Mars. Some of their gut biome came from the frozen remains of their woolly rhino ancestors, some from musk oxen and the rest from rhinos still existing in the 2800s.

Mammoths preferred temperate forests on Mars - but woolly rhinos - like musk oxen and other ice age species, favoured tundra and taiga environments in the regions around the southern ice cap and on the slopes of the Tharsis Mountains. Successful ecosystems also appeared in the northern basin. However even though they flourished in some areas, they were all eventually drowned as the great northern ocean filled.

CHAPTER 13
RIVERS

As Mars' surface became wetter due to comet impact and the melting of ancient ice, trickles of water turned into streams, streams into rivers, and rivers into estuaries. Great deluges tore their way across Mars' deserts and through mountain ranges, creating three huge rivers where just dry land had existed before. In some regions lakes appeared as melting ice filled craters; in other areas resuscitated river courses followed the same routes they'd taken billions of years before.

Once drainage patterns were established and the water had cleared, the seeding process could begin. The temperate and cold climate rivers on Earth which could have been used as life sources for seeding included the Amur in China; the Yukon in Alaska; the Mackenzie in Canada; the rivers Yenisey, Lena, Kolyma and Ob in Siberia; and the Rhine in Europe.

The shortest and steepest of Mars' new rivers had its source on the northern part of the extinct volcanic mountains of the Tharsis bulge, and emptied steeply eastward into the colossal Valles Marineris canyon. The colonists named this river, the "Einstein".

The second river, the "Newton", began in the southern part of the Tharsis bulge and flowed south before turning east through the Argyre Planitia depression, from which it flowed north into the ocean. Both the Einstein and the Newton emptied into the Chryse Planitia area of the northern ocean.

The last of the three great rivers, the "Darwin", was sourced from multiple small rivers which flowed into the Hellas Planitia basin. A huge river then flowed north from there into the Isidia Planitia region of the ocean.

The different nature of the three rivers; the Einstein being short and steep, whilst the Newton and the Darwin were long with lakes on their courses; meant that the rivers used as sources for life on Earth needed to share at least some of the same geographical and climactic characteristics as their equivalents on Mars.

The Yukon in Alaska was chosen as the source of life for the Einstein because of its steepness and the fact that almost all of its drainage basin was confined to mountainous terrain. The Mackenzie in Canada was initially targeted as a life source for the Newton because it also had lakes on its course, and some of its tributaries came from mountainous regions. However, it was later shown to have a very similar ecosystem to the Yukon, so the Amur in China was finally selected instead. The Amur had a unique ecosystem as well as having the lake Zeyskoe on one of its tributaries. The Lena in Siberia was chosen as the life source for the Darwin as it had the unique Lake Baikal ecosystem on its course.

The seeding of life onto land and into rivers both overlapped and complimented each other as many of the species to be seeded were common to both schemes. The first rivers had life added to them fifty years after land seeding was started, which helped to keep the two seeding schemes in sync with each other.

The difficulty with seeding rivers is that they change their nature between source and sea. This meant that river seeding had to be split into the four ecosystems on their courses; mountain streams, lakes, main channel and estuary seeding. These were dealt with as separate biomes and seeded one-by-one. Nevertheless, these Earth ecosystems do intermingle as the flow of water takes life downstream, and many life forms have strategies which help them travel back upstream as well.

13.1 The River Yukon
The Einstein was the most difficult of Mars' rivers to seed due to the enormous Mendeleev waterfalls which cut the river short where it entered the Valles Marineris canyon. The terraforming biologists could have used one of the rivers rising in the Himalayas (Ganga, Brahmaputra and Indus), but their main channels and lower reaches were too warm to be considered as seeding sources for the Einstein.

No cold climate river on Earth was split in two by such a dramatic barrier. However, the Yukon was cold enough to have similar climatic extremes to that expected on Mars and there were two interruptions in its flow, one at the Five Finger rapids and the other at the Rink rapids. The Yukon also had numerous lakes and small streams amongst its tributaries which was what was expected with the Einstein on the slopes of the Tharsis Bulge.

The catchment of the glacier-fed Yukon River covers a vast area of Alaska and Canada. Its topography is highly variable from extremely high mountains in the east to rolling hills in the west. It flows over unstable screes and gravels, so that the river carries a great deal of suspended sediment in its waters.

The land seeding sources which overlap with the Yukon include Mount St Helens and the Low Arctic Taiga. Yet, due to the height of the Tharsis Mountains and the low temperatures expected there, only Low Arctic Taiga was really appropriate for land seeding. Hence some supplementary land seeding took place alongside the Einstein using Low Arctic Taiga lifeforms.

The upper parts of the river Einstein were expected to be similar to the Yukon with lakes and screes, but below the Mendeleev falls the river was predicted to take the form of a fiord with a mixture of saline oceanic and fresh river water. The fiord then entered the Isidis Planitia section of the northern ocean as a part of the estuary it shared with the Newton. This meant that the seeding of the Einstein was divided into three parts; with life from the Yukon being used in the upper part, oceanic life being added in the fiord and life from the river Amur in the estuary. The estuary was seeded at the same time as the Newton estuary with life from the river Amur.

Microorganisms from the Yukon could be frozen because they had evolved to survive severe winter temperatures over millions of years. The types of microorganisms in the Yukon's gravels and muds were controlled by the nature of the underlying rocks, which meant that some seeding had to be sensitive to the underlying geology. So, if a mud sample came from a section on the Yukon with a basalt substrate, it would have a better chance of surviving if it was added to a basaltic section of the Einstein. The seeding was therefore addressed in 7 stages and, within each stage, consideration was given to the site of seeding – whether it was a stream, lake or river.

Stage	Streams	Lakes	Rivers
1	Mud and gravel microorganisms	Mud and gravel microorganisms	Mud and gravel microorganisms
2	Diatoms	Copepods, rotifers, cladocera and ostracods	Diatoms
3	Insects, annelids, worms, gastropods and crustaceans	Clams, gastropods, aquatic worms and flatworms	Insects, annelids, worms, gastropods and crustaceans
4	Dragonflies and mayflies	Dragonflies and mayflies	Flies, mayflies and arachnids
5	Lamprey, grayling and trout	Trout, grayling, burbot, inconnu, pike and whitefish	Lamprey, grayling, burbot, inconnu, pike, trout and whitefish
6	Wood frog	Wood frog	
7	Ducks, beaver and geese	Ducks, beaver and geese	Ducks, beaver and geese

One notable exception to the seeding of the Einstein was Salmon, an important contributor to life on the Yukon. It was recognised that this fish would not survive the drop down the Mendeleev falls and, even if it could, it would not be able to swim back up them.

13.2 The River Amur

The Newton was far more of a conventional river than the Einstein as it had a normal watercourse without a dramatic waterfall barrier on its course. The Newton's flow started in the drainage of the southern part of the Tharsis bulge and then looped around and drained into the Argyre Planitia depression where it created a large lake. Its path then continued north and drained into the Chryse Planitia estuary.

This more typical course and gradual descent meant that the river Amur was a good choice as a source for life. Additionally, the Amur was highly diverse with many life forms and overlapped nicely with the land seeding done as part of the Chinese Temperate Broadleaved Forests initiative.

Both Earth's Amur and Mars' Newton were affected by dramatic flooding, as the Amur was influenced by the monsoon which caused heavy floods and the Newton was affected by its high mountainous source and intense seasonal rains. The Amur suffered damage to its bottom communities which were often washed out, and similar patterns affected the Newton. Mars' Newton was also affected by landslides and avalanches which caused flash floods and gravel avalanches.

There were two issues which caused a great deal of discussion during the Amur – Newton seeding. The first was the presence of biting midges in the Amur which the biologists wanted to introduce to the Newton. However, understandably, most colonists did not want biting insects in their rivers as it could have made some parts of the Newton's catchment unendurable. It was therefore agreed to introduce an additional species of non-biting midge to ensure there was sufficient diversity to support fish and water birds.

The second issue related to the need for a top river predator. One of the four species of sturgeons is the Kaluga, a predatory salmon eater which migrated between the main river channel and the sea. Concerns were raised that the number of Kaluga in the Amur were always low which meant that it would probably not fill the role of a top predator in the Newton.

For that reason river dolphins were suggested as a possible solution. Yet, by 2500, all river dolphin species were extinct. It was possible to bring one back, but no river dolphin had had blubber, which meant that they preferred warm waters and were unlikely to survive the Newton's cold conditions. Whilst it might have been possible to mix the genes of a river dolphin with those of a cold water marine species the biologists preferred to use a DNA from an real species rather than attempt to create one which had a low chance of success.

There was discussion about using the last remaining freshwater seal, the Lake Baikal Seal. Some biologists thought that these seals would be the best solution, but it was countered by other biologists that these were expected to be needed in the even larger Hellas inland sea which was part of the river Darwin seeding plan. Furthermore, neither dolphins nor lake seals had evolved to be able to swim over rapids and cascades.

The focus of the discussions then turned to otters which were present in China. The Oriental small-clawed otter was mainly restricted to southern China and preferred warmer conditions; so it was agreed that the Eurasian otter, found in the Amur, was to be selected as the key predator for the Newton. So another seven stage seeding campaign took place:

Stage	Streams	Lakes	Rivers	Estuary
1	Mud and gravel microorganisms	Mud and gravel microorganisms	Mud and gravel microorganisms	Mud and gravel microorganisms
2	Amphipods and phytoplankton	Amphipods and phytoplankton	Amphipods and phytoplankton	Phytoplankton including diatoms and algae
3	Bivalves, worms and gastropods	Bivalves, worms and gastropods	Bivalves, worms and gastropods	Zooplankton
4	Caddisflies, mayflies and stoneflies	Non-biting midges and caddisflies	Caddisflies, beetles, non-biting midges, flies, stoneflies, dragonflies, mayflies, damselflies and crustaceans	Bivalves, worms and gastropods
5	Grayling	Carp, catfish and perch	Carp, perch, bream, catfish, grayling, barbell and minnows	Carp and perch
6	Salmon and trout	Pike and snakehead	Salmon, trout, sturgeon, pike and kaluga	Paddlefish, sturgeon and kaluga
7	Otters	Otters	Otters	Otters

There were a number of issues with the Amur's estuarine ecosystem. These included its unusually warm waters in the cold Okhotsk Sea, although this did work nicely with the warm Newton due to geothermal inflows and the cold northern ocean on Mars. There was discussion before the estuarine seeding began about using a completely different assemblage of life forms from a colder estuary. In fact, the Kolyma estuary was suggested for this. The Ob and Yenisey had estuaries which were long flooded glacial valleys and were deemed too different to be considered. However, once optimum temperatures had been planned and the wind towers were in operation warming Mars' atmosphere, the similarity between the temperature ranges of the Amur and the Newton made this unnecessary.

The second issue was the loss of the paddlefish which had become extinct around the world in the 2100s. These were important as they were key zooplankton filter feeders which could move easily between estuary, rivers and lakes, making protein for top predators like otters, cats and birds.

There had originally been five species but the last two of these, being the Chinese and the American paddlefish - had been overfished and polluted out of existence long before they were needed for Mars. This necessitated work to try and bring them back from extinction using fragmentary DNA samples stored in formaldehyde in museums and was not as easy as was hoped. The principal problems were the damaged state of the DNA and the small number of samples available. This resulted in the need to use DNA from both Chinese and American paddlefish to create a fish which was a blend of the two. Then this had to be further altered to create species diversity so as to avoid inbreeding and hereditary diseases. Seeding the Newton necessitated six attempts at introducing paddlefish before a viable breeding population was established; the previous five attempts had seen all the paddlefish dying shortly after introduction.

Another issue which was important to the health of the Newton's ecosystem was the transport of resources back from the ocean to the land by migratory fish. These fish were included in as additional stage 8, and were an important resource return system in the river Amur. They were expected to behave in the same way in the river Newton. Migratory fish included various species of salmon, trout, paddlefish, sturgeon and kaluga. All of these species require an ample supply of oceanic food including zooplankton, insects, and shrimp. This meant that river and oceanic seeding had to take place at the same time and the northern ocean had to be rich in invertebrates before migratory fish could be introduced to the Newton.

13.3 The River Lena

There were a number of similarities between the Lena in Siberia and the Darwin on Mars as both had large lakes interrupting their courses, both had numerous mountainous inflows feeding their respective lake, both had mountainous upper river courses and both had deltas at their contacts with the sea.

In spite of these similarities, before seeding began, the assumptions had to be tested - where it was possible - under laboratory conditions. Some things were reasonably certain, for example the Hellas Planitia basin is over 4,000 meters deep and so was likely to need deep cold fresh water lifeforms. Lake Baikal is Earth's coldest deep water biome and was therefore an obvious choice as a seeding source.

Other things were less obvious. For example, Lake Baikal is a well-oxygenated lake without any water stratification. The Hellas basin was considerably deeper and was likely to have numerous inflows from the surrounding mountains, meaning it would probably also be well-oxygenated. The difficulty with Hellas was the implication that due to its depth and the presence of salts in the surrounding rocks, it would start to accumulate these chemicals which could stratify into a deep brine layer overlain by fresh water. That said, it was countered that it was possible that the incoming rivers would prevent this happening at Hellas.

From studies of the geology and landforms around the Hellas basin it was expected that Hellas would fill to a depth of about 5,000 meters before violently cutting a way either north or northeast to the northern ocean. This outflow was expected to be both very powerful and to carry enormous quantities of rock and sand to form a delta within a short space of time. The size of this delta would depend on the route the river took, as the north east-escape route would allow the delta to develop in shallower seas whereas the northern route emptied into much deeper waters.

This process was also expected to lower the water level in the Hellas basin and to cut a deep canyon through the surrounding mountains. This is, in fact, what happened when the basin filled to its maximum depth and extent in the 2440s and seemed to use both northern and north-eastern routes to the sea before dramatically rending a route through interconnecting lava tubes on the north east route. It quickly tore the roofs from these tubes and used them as a shortcut into the ocean, thus reducing the extent of the Hellas inland sea to half its previous size.

In the early decades of its existence the Hellas basin was subject to both fierce inflows and equally fierce outflows. Massive amounts of frozen water were arriving in Mars' upper atmosphere producing monsoons which lasted for months and which caused powerful and fast-moving new-born rivers to begin filling Hellas. This ensured that no stratification took place even though considerable amounts of salt were being washed into the depression. Later, the sudden outflow, through the newly created Darwin River, also helped to prevent stratification. However in the 2600s some indication of stratification was noted and a number of more salt-tolerant species were found living below the salinity horizon. Hellas' upper layers were highly oxygenated due to the number of rivers entering the basin, allowing many large Baikal amphipods to survive and do as well in Hellas as they had in Baikal.

After the partial emptying of Hellas, a number of islands appeared around its edges. This echoed the situation on Lake Baikal and ensured a safe refuge for the lake's top predator, the Baikal seal. These were introduced to Hellas 10 years before the first large predator was introduced in the land seeding zones.

The main channels of the Lena and the Darwin both cut their way through considerably mountainous regions, accumulating gravel and sand below their rapidly moving waters. This increased the chances of microorganisms and invertebrates from the Lena being able to prosper in the Darwin. Spring floods became a regular event, scouring the river bed and developing large flood plains in the Darwin's lower reaches.

The seeding of life into the Darwin was one of the most complex projects undertaken as part of the terraforming initiative. The delta needed tundra additions and the main channel needed taiga additions, making the river Darwin a three-source zone. The seeding campaign consisted of 7 distinct stages:

Stage	Lake / Basin	River and flood plain	Delta
1	Mud and sand microorganisms	Sand and gravel microorganisms	Mud and sand microorganisms
2	Phytoplankton including diatoms and algae	Worms and leeches	Phytoplankton including diatoms, algae, dinoflagellates and nanoflagellates
3	Zooplankton including copepods	Insects including blackflies, grass-flies and non-biting midges	Zooplankton including rotifers, arthropods, copepods, water-fleas and protozoans
4	Sponges	Siberian wood frog	Fish including Arctic lamprey, burbot and whitefish
5	Amphipods, crustaceans, snails, bivalves and worms	Salmon species, grayling, whitefish, burbot, roach, dace, carp, minnows, gudgeon, bullheads and perch	Salmon and sturgeon
6	Fish including sculpins, oilfish, whitefish and grayling	Salmon and sturgeon	Many bird species
7	Seals	Pike	Buzzards, merlins, skuas, snowy owls and peregrine falcons

The delta was particularly complex as shoreline tundra seeding had to be performed because the continental tundra ecosystem which had previously been undertaken was lacking in maritime species. Additionally, much of the tundra bird life which needed to be added to the delta would need to migrate in the winter, probably to the warmer Hellas and Argyre basins. During Mars' winters the delta would freeze for as long as eight months, meaning that only the most cold-tolerant of species would survive. It was nonetheless recognised that the Lena's delta on Earth froze each year, albeit for a lesser duration, giving Lena's species a chance of survival on Mars.

Much of the tundra ecosystem seeded on the seed line to the south had to be repeated on the Darwin's freshly built delta. A number of predator species needed to be excluded, such as the arctic fox, in order that the bird life would have a chance to establish itself. Predatory birds also needed considerable delay before being introduced so that prey numbers could grow.

After the initial tundra seeding, birds needed to be added. The addition of birds included the introduction of many species such as bottom feeders like swans, geese and ducks; insect feeders such as plovers and sandpipers; and fish and crustacean eaters like gulls. In all over 300 bird species were added, but roughly half of them failed to survive.

On Earth the Lena's delta was subject to a ten meter tidal range. On Mars the shore of the northern ocean was expected to have a tidal range of just a meter or two. This did cause a problem for many species as the ecosystem was less productive. Serious discussion took place at the time about bringing a new and larger moon to Mars to increase the tidal range. This, however, wasn't followed through as the introduction of such a moon was outside human capability at the time.

13.4 River Summary

In theory rivers should be easy to seed, but they are not. This is due to shortages of nutrients and the failure of microorganisms that had been introduced at the start of the process. Both of these resources were at risk of being washed out to sea by floods or simply due to the normal flow of water. Species can fail to survive because of the chemical balance of the water, toxins in the environment or the absence of a key food item. Mars' low gravity did sometimes prevent survival due to water foaming, slow water movement or insufficient oxygenation.

Nevertheless, some species did survive, but they were not always the ones that were predicted to be successful. In fact, some species which did well on Earth failed to survive on Mars and vice versa. Being introduced from Earth to a different planet forced life forms to behave in changed ways and this increased the difficulty in predicting outcomes during terraforming. The most challenging part was analysing which of the thousands of microorganisms in river muds were critical and finding out if they survived and why.

There were patterns in the type of life form which survived in Martian rivers. Much of these patterns were driven by water chemistry, speed and temperature. In future terraforming projects these key factors could be used to drive decisions about what to introduce, when and where.

CHAPTER 14
OCEANS

By 2440 the northern basin had started to fill, creating Mars' first ocean. Although colonists had to stay well away from water due to frequent land slips and tsunamis, there was great interest in adding life to Mars' marine environments. However the northern depression was so turbulent that light could not penetrate it dark waters.

By 2497 northern ocean had cleared sufficiently to allow life to be seeded at long last. Samples were collected from many oceanic sources using a methodology which entailed dropping a filter basket over the side of a ship so that it descended several hundred metres. This basket had a mesh on its sides and bottom sufficiently fine to collect plankton, but was lidded with a courser mesh to keep out larger objects. This prevented the boxes from collecting the adult stages of the species, which were represented in the larval plankton.

The filter basket was then slowly pulled back to the surface allowing it to collect as many five thousand species and a million individual microscopic planktonic life forms. When it reached the surface it was placed into a water retaining box which was hoisted on board ship and was then flash-frozen. This approach was taken as there were simply too many forms of oceanic life to 3D print at this time.

Hundreds of these one litre oceanic life slices were collected from each source zone which included many of Earth's most biodiverse regions. The first three collection zones to be sampled were chosen to match the sampled rivers. So, the Yukon was matched by Alaskan, the Amur by Sea of Okhotsk and the Lena by Arctic oceanic life samples. This was intended to ensure that the land, river and ocean all worked as a part of the same super-scale ecosystem.

For example, the Arctic ocean provided the food and other resources needed by migratory salmon, which swam up the Lena to lay eggs in the streams and were then caught by bears and otters which fertilised the Taiga forests.

Additionally, so as to avoid missing any bottom dwelling micro-organisms, mud samples were collected at each site so that bacteria, viruses, fungi and protozoans were included in the seeding on Mars.

The way these species were seeded was that each oceanic life slice was simply dropped from the air into an appropriate part of Mars' northern ocean. These frozen life slices would then melt and release their plankton into the upper layers of Mars' ocean. The frozen mud slides were dropped inside a weighted frame. Once on the sea floor the sample would melt and release its invertebrates into the mud. The appropriateness was determined by local oceanic salinity, clarity, toxicity, oxygenation and nutrient content. Cliff stability and substrate composition were also important.

Life slices sourced from the same place were seeded into Mars' ocean at regular intervals, depending on local flying conditions. This usually meant seeding a number of regions of the ocean during each flight.

An early estimate of seeding success was that about one present of each slice survived, which meant one hundred of the ten thousand life forms survived the freezing and seeding processes. Phytoplankton such as algae were the primary early forms of life to prosper as their primary need was sunlight but, as the years passed and microscopic plants proliferated, zooplankton began to survive the seeding process as well.

The next batch of seeding survivors included bacteria, archaea and fungi. These were followed by the eggs and larval forms of sea urchins, crustaceans, jellyfish, corals, sponges, starfish, marine worms, squid, octopus and finally fish. Some life forms failed to survive the seeding process and had to be 3D printed, as did larger fish and sharks. After 2700 sea birds and sea eagles were introduced, followed by seals, sea otters and walruses during the 2800s.

These apex oceanic predators were critical to preventing disease and controlling numbers lower down the food chain. Without these predators, there would have been population explosions followed by disease epidemics, both of which would have significantly impeded the seeding of Mars' ocean.

Mars' northern ocean was terraformed far faster than the land. This was in a large part due to the in washing of nutrients such as iron from the land by rain, which was then carried in streams and rivers and resulted in a significant initial overgrowth of phytoplankton but meant that zooplankton prospered far quicker than had been expected.

Whales and other cetaceans were a problem, however as due to the foaming caused by Mars' low gravity, there was a risk that they could drown in stormy weather. The first dolphin species was introduced in 2627 but their breeding rate was so low that no others were introduced until after 2800 when mega storms became less of a problem. After this, resuscitated approximations of cold water whales including beluga, orca and narwhals were bought back from extinction and seeded into tanks with robotic mothers for the first parts of their lives.

Water ice quickly replaced frozen carbon dioxide at both the North and South Poles. The North Pole ice cap formed as floating sea ice held in place by the gyration and tidal flow of the northern ocean. This permanent sea ice and the Elysium Island were seeded with various species of penguins in 2670 and, in absence of land predators such as polar bears they prospered.

Elysium Island in time became Mars' greatest sea bird colony with millions of birds including Great Auks, Terns, Puffins, Gannets, Petrels, Skuas, Fulmars, Shearwaters, Guillemots, Geese and Albatrosses. Elysium also played host to huge colonies of breeding seals and walruses.

14.1 Southern Alaska - keystone Kelp

The two key locations used for the sampling of Alaskan marine ecosystems to compliment the Yukon river collections were the shallow water Glacier Bay situated south east of Anchorage and the deep water Ikatan Bay in the Aleutian Islands. The species gathered from these two locations were seeded in Mars' great rift, the Valles Marineris canyon, and sections of the Chryse and Acidalia Planitias. The season in which the collection was made from Earth and the seeding on Mars completed had to match to improve chances of survival. Other factors which needed to match as closely as possible were water temperature, salinity and chemistry.

Glacier Bay
Like all the oceanic sampling sites, the primary aim was to sample phytoplankton and then zooplankton. This helped build an ecosystem from its most simple forms upwards. The sampling was mainly conducted in early spring when kelp plankton was at its maximum. This meant that marine plants could be included in the samples sent to Mars. The spring sampling also included the larval forms of krill, copepods, starfish, sea urchins, snails, worms, clams and fish such as herring, mackerel, cod and skates.

The final parts of this ecosystem included sea otters, seals and birds; but these had to be created in the laboratory by recreating their DNA by re-sequencing machines turned computer data into DNA. The recreated DNA was then injected into animal cells which had been 3D printed. These were animated by electric stimulus, then incubated in artificial placentas and finally raised by robotic mothers.

Ikatan Bay
The deeper ocean was sampled in the Aleutian Islands by dropping the mesh basket far deeper than they had been deployed in Glacier Bay. This meant that the majority of the collected life was from below a depth of 500 meters. This worked well for the majority of species apart from phytoplankton which had to be collected from nearer the surface due to their need for sunlight. The timing of most of the collections were set to coincide with the coral spawning and

had to be predicted by environmental factors such as the lunar cycle, water temperature and day length.

Amongst the thousands of species collected in the plankton soft, gorgonian, cup and black corals were well represented, as were sea pens and sea whips. The larval forms of octopuses, squid, starfish, crabs, scallops, sea anemones, basket stars, crinoids and shrimp were also all present in large numbers. Whilst some larval forms of fish such as rockfish, mackerel and flatfish were present, survival after freezing was poor and they had to be DNA re-sequenced to ensure survival in sufficient numbers to succeed on Mars.

More advanced top predator sea birds such as kittiwakes and albatrosses also needed to be DNA re-sequenced and introduced gradually after they had been taught the rules of survival by robotic mothers. This was also true of some species of seal which complimented those collected at Glacier Bay.

Steller's sea cow was a later addition to the Ikatan Bay ecosystem. These large aquatic kelp eating mammals had once lived in around the Komandor Islands in the Aleutian chain until they had been hunted to extinction in the 1770s. Bringing them back from extinction was achieved by collecting fragmented DNA from their bones and comparing it with the genetic code of their existing relative, the Dugong. This allowed the artificially intelligent resequencing machines to recreate a close approximation of sea cow genetic code and to then use this to create the sea cows for Mars. These were reared in lagoons by robotic parents and then released into the Valles Marineris canyon area when they reached maturity.

14.2 Sea of Okhotsk

The Sea of Okhotsk was used as a sample source for marine ecosystems to complement the river Amur collection and therefore the seeding of the river Newton. The Siberian Okhotsk Sea has an average depth of just less than 900 meters and a maximum of a little over 3,000 meters. Winter temperatures drop to minus 15 degrees Celsius and much of the ocean can be iced over. Summer temperatures normally rise to 20 degrees Celsius but ice can still be seen in the sea as late as May.

Dzhugdzur

Life from the marine portion of the Dzhugdzur reserve was used to seed the Chryse and Acidalia Planitias seas on Mars. With its relative proximity to the Amur River and to the taiga land ecosystem, it was a good source for marine species. This meant that with supplementary seeding of Low Arctic Taiga species into the nearby land regions and the seeded river Newton emptying in the same sea area, the three ecosystems complimented each other nicely.

The Dzhugdzur is a particularly diverse ecosystem rich in both in phytoplankton and zooplankton making it a good candidate as a seeding source. At the right time of year its plankton also included the spores of kelp and other seaweeds which were a useful addition to the ecosystem on Mars even if only the coastal portions of the target Planitias seas were shallow enough for sunlight to penetrate. Other species forming parts of the plankton included the larval forms of squid, octopuses, crayfish, mussels, crabs, sea urchins, polyps, shrimp, snails, amphipods, barnacles, sponges and worms.

In addition to these invertebrates, the region was very rich in fish such as herring, pollack, flounder, cod, turbot, capelin and smelt. These as well as sea birds, seals, sea lions and later whales had to be DNA re-sequenced rather than simply collected and seeded as part of the plankton.

14.3 Arctic

To match the river Lena and to provide a cold water ecosystem for the northern ocean off the mouth of the river Darwin, the Arctic Ocean was used as a source. However, the lack of ice for most of the year meant that it was not a perfect source for Mars as its temperature had risen over the centuries. As a result of this warming, the water temperature was well above the freezing point of sea water for nine months of the year. Therefore, part of the ecosystem could come from the arctic, but the rest would have to come from elsewhere.

Open water

Sampling in the Arctic Ocean took place north of Greenland in the Eurasian Basin which is in parts over 5000 meters deep and was where the last summer ice cover had existed. This increased the chances of arctic species being cold and depth tolerant and hence able to survive Mars' deep cold ocean.

Both deep and shallow parts of this portion of the Arctic Ocean were rich in Phytoplankton, zooplankton and seabed microorganisms, making it a good source for the cornerstone ecosystem ingredients. Thereafter successive ice slice seeding added crabs, shrimp, shellfish, brittle stars, starfish, sea urchins, snails, worms, jellyfish, copepods and krill, followed by more complex consumers such as squid, cod, herring, redfish, halibut and plaice. Following this, fully marine predators such as rays and sharks were introduced before the addition of seals, walrus and birds, which were subsequently added to keep the ecosystem healthy by removing sick or damaged specimens. The final top predators, whales, were not added until the increasing air pressure, brought about by the introduction of the inert gasses in Mars' atmosphere, reduced oceanic spray and foaming.

Under Ice
The last summer ice had disappeared from the Arctic Ocean in the 21st century and winter ice was thin and unstable. This meant that many key ecosystem elements like ringed seals and polar bears had been gone for hundreds of years. However, there would be thick and stable floating ice in Mars' northern ocean so species which could cope with life in freezing conditions was needed. The only ecosystem on Earth which still included floating ice was Antarctica where sections of this southern continent still functioned as they had done centuries before.

This argument, as well as one concerning the introduction of penguins, came to dominate discussion amongst the terraformers in the 26th century. The problem was that they and Mars' populace at large wanted penguins. Penguins and polar bears were not compatible; they had to choose one or the other and couldn't have both.

It was finally decided to collect life from that portion of the Antarctic which still had summer sea ice. This was in its design the simplest ecosystem transferred to Mars. The key fundamental elements were phytoplankton, herbivorous and carnivorous zooplankton, bacteria, protozoa, all of which were highly tolerant of freezing. Following which, the collection of invertebrates, such as the cornerstone krill - which served as a key food source in this ecosystem – was completed. This was accompanied by the collection of molluscs, starfish, copepods and Jellyfish.

Key consumers seeded after this included fish such as Antarctic Cod and the Mackerel icefish, and squid which make up large portions of the marine Antarctic ecosystem on Earth.

Thereafter additional species of seal were seeded such as elephant, crab eater and finally predatory leopard seals. These complimented the seals and walruses seeded as parts of other marine seeding programmes.

The next stage of the seeding included seven species of penguin which were seeded onto the floating North Pole ice sheet and onto Elysium Island where they joined huge colonies of seabirds such as snow petrels, albatross and predatory skuas. Finally additional species of whale were added to this part of Mars' ocean after 2800.

14.4 Additional Marine Seeding

Marine seeding on Mars was as much a scattergun approach to introducing life on a new planet as were the land or river seeding projects. As the centuries have passed and we have learnt more about terraforming, we have become better at choosing the right ecosystem to collect and the right place to seed it. But in the 2400s we didn't understand how closely chemistry, temperature and moisture determined what would survive and what wouldn't.

In the 2680s the marine extinction rate was analysed and found to be far higher than expected. This was believed to be due to the different chemistry and salinity of Mars' ocean and its dissimilarity from any ocean on Earth. However, Mars' ocean was very rich in minerals and nutrients and those species which survived did very well.

Therefore, new analyses were made of the relationships between ecosystems and environments in order to be more precise about the correct species to introduce. In the 2700s new ice slices, which included additional kelp species, were seeded into Mars' ocean from the coast of Norway, Iceland, South America, Eastern New Zealand and Eastern Japan. The larval forms of cold water corals and their ecosystem companions were sourced from the waters off Scotland, Ireland, Norway and Greenland. Together these kelps and corals helped to support their ecosystems and more of these survived than in the original seeding.

14.5 Oceans Summary
The great depth of Mars' Ocean and the concentration of heavy water at its deepest portions should have made life in the deep unfriendly to life. However, the concentrations of minerals and nutrients being accumulated at these depths made this environment very beneficial to those species which could endure it.

And so the depths became as friendly to life as the rivers, plains and mountains; and the availability of nutrients in the runoff from the continent ensured rapid colonisation, proliferation and adaption in Mars' ocean. Cyclic movement of ocean currents and the influence of low tides caused by Mars' two small moons ensured reasonable oxygen supply throughout the ocean. This was complemented by the influence of the three well oxygenated rivers pumping water into the fringes of the sea.

However, questions remain about the first human attempt at terraforming; are Mars' small moons enough to maintain ocean currents and keep oxygenation going in the long term, or should a new and larger moon be introduced to the planet? And if the first life to evolve was that living in oceans on Earth, should seas be terraformed first on other planets in the future?

CHAPTER 15
CONCLUSION

Life is a major tool enabling humans to take a dead planet and turn it into a good home for humans. Having living things on a planet's surface is the best way to convert inorganic minerals into organic compounds people can eat or use.

Yet should there be a standard procedure for terraforming so that each step is clearly pre-planned whatever the nature of the planet? Should some environments take precedence over others? Should oceans be terraformed first using a 'life follows water' approach, followed by rivers and then the land? Or should it all be done in the way it was done on Mars? Or is there a better way to schedule that we haven't discovered yet? For example, should we use mineralogy to determine which sites to terraform first?

Additionally should the availability of organic proteins be used as a guide to determine the sequence of seeding within an ecosystem? So, whilst it is easy to assume that lemmings need to be seeded into the tundra environment before owls, are there other sources of proteins the owls need which might make lemmings less critical? And, if lemmings did not prosper in the tundra, what other species could be used in their place if owls are still needed as top flying predators?

On Mars the seeding was determined by technological limitations to travel and flight. Rain fell first on land close to the colony and so that was where terraforming began. If distance were no longer a problem, how should terraforming be planned, scheduled and performed?

Each planet has a maximal DNA data storage potential which means that the total number of species and members of each species add up to a total terabyte planetary memory capacity.

Thus if terraforming is planned for a new planet, that planet's DNA data storage capacity can be calculated from its size, temperature, water availability and mineralogy. This would then allow the terraformers to plan their ecosystem introductions with this maximum in mind. As a result, those involved in the terraforming of a planet would be able to determine how many ecosystems to introduce before the maximum is approached. This also meant that they would be able to prevent large scale extinctions from occurring as had happened on Mars. Importantly, the total storage has to include all the humans living on that planet when fully colonised and must to take account of the growth of the maximum over the centuries as organic nutrients become increasingly available.

Mining on Mars has changed over the centuries from an industry needing many people to work in the mines to an industry driven by an automatic process with very few people involved. Nano machines have made people superfluous in the process as they can extract the minerals from deep sources, collect them into silos and organise the shipping. Humans are only needed to sell the minerals, monitor the total shipped capacity and ensure pollution does not damage Mars' environment.

Mars's atmosphere remains dynamic in nature. Firstly the upper layers are continually stripped by the solar wind; secondly, carbon dioxide has to be continually replenished to power photosynthesis; and thirdly, the composition has to be continually adjusted to maximise atmospheric pressure and breathability. Water on Mars also has to be dynamically replenished. The solar wind is again at fault as it strips a proportion each year; however there is also a loss as some water sinks into the planet's crust.

If Mars were to be terraformed today as a newly discovered planet a number of things would be done differently. The annual rotation would be speeded up to make a year 365 days of 24 hours each. A new and larger moon would be added in place of phobos and deimos to give the ocean larger tides. This would speed ocean currents meaning the waters would be better oxygenated. And the planet's core would be restarted to warm the planet, regulate the atmosphere by outgassing water, carbon and oxygen and develop a magnetic field to protect the planet's life and atmosphere.

But that, as they say, is another story.

Milton Keynes UK
Ingram Content Group UK Ltd.
UKHW021514170524
442874UK00033B/254

9 780995 674127